Analog Circuits and Signal Processing

Series Editors

Mohammed Ismail, Wayne State University
Dublin, OH, USA

Mohamad Sawan, School of Engineering, Westlake University
Hangzhou, Zhejiang, China

The *Analog Circuits and Signal Processing* book series, formerly known as the *Kluwer International Series in Engineering and Computer Science*, is a high level academic and professional series publishing research on the design and applications of analog integrated circuits and signal processing circuits and systems. Typically per year we publish between 5–15 research monographs, professional books, handbooks, and edited volumes with worldwide distribution to engineers, researchers, educators, and libraries. The book series promotes and expedites the dissemination of new research results and tutorial views in the analog field. There is an exciting and large volume of research activity in the field worldwide. Researchers are striving to bridge the gap between classical analog work and recent advances in very large scale integration (VLSI) technologies with improved analog capabilities. Analog VLSI has been recognized as a major technology for future information processing. Analog work is showing signs of dramatic changes with emphasis on interdisciplinary research efforts combining device/circuit/technology issues. Consequently, new design concepts, strategies and design tools are being unveiled.

Topics of interest include: Analog Interface Circuits and Systems; Data converters; Active-RC, switched-capacitor and continuous-time integrated filters; Mixed analog/digital VLSI; Simulation and modeling, mixed-mode simulation; Analog nonlinear and computational circuits and signal processing; Analog Artificial Neural Networks/Artificial Intelligence; Current-mode Signal Processing; Computer-Aided Design (CAD) tools; Analog Design in emerging technologies (Scalable CMOS, BiCMOS, GaAs, heterojunction and floating gate technologies, etc.); Analog Design for Test; Integrated sensors and actuators; Analog Design Automation/Knowledge-based Systems; Analog VLSI cell libraries; Analog product development; RF Front ends, Wireless communications and Microwave Circuits; Analog behavioral modeling, Analog HDL.

Byunghun Lee • Hyun-Su Lee •
Junhyuck Lee • Hyung-Min Lee

Energy-management Integrated Circuit Design for Wireless Power and Data Transfer Applications

Springer

Byunghun Lee
Department of Biomedical Engineering and
Department of Electronic Engineering
Hanyang University
Seoul, Korea (Republic of)

Junhyuck Lee
Department of Electronic Engineering
Hanyang University
Seoul, Korea (Republic of)

Hyun-Su Lee
School of Electrical Engineering
Korea University
Seoul, Korea (Republic of)

Department of Biomedical Engineering
Hanyang University
Seoul, Korea (Republic of)

Hyung-Min Lee
School of Electrical Engineering
Korea University
Seoul, Korea (Republic of)

ISSN 1872-082X   ISSN 2197-1854   (electronic)
Analog Circuits and Signal Processing
ISBN 978-3-032-00744-5   ISBN 978-3-032-00745-2   (eBook)
https://doi.org/10.1007/978-3-032-00745-2

This Springer imprint is published by the registered company Springer Nature Switzerland AG
The registered company address is: Gewerbestrasse 11, 6330 Cham, Switzerland

If disposing of this product, please recycle the paper.

Contents

Chapter 1
Introduction

1.1 Introduction

Wireless power transfer (WPT) is one of the crucial technologies that supplies power in various applications with indirect contact between the energy source and wireless devices. There are wireless power applications with diverse power requirements, from nanowatts in wireless sensors and radio frequency identification (RFID) tags, milliwatts in near-field communication (NFC), watts in mobile electronics, to kilowatts in electric vehicles. Figure 1.1 illustrates some of the state-of-the-art applications for WPT. High power transfer efficiency (PTE), robustness against nearby objects and coil misalignments, and an extended power transfer range are highly desired in all these applications.

For example, wireless charging for smartphones is widely used in daily life, and applications such as wearable devices and portable electronics also demand high performance and power efficiency. These wireless devices require more power to handle more functions on a larger scale under variable environmental and loading conditions. Particularly, when complex data processing or communication with various sensors is needed, the power level is inherently high and less dependent on the power consumption of internal circuits. In most cases, supplying these devices with primary batteries may not be an option due to their large volume, limited lifetime, difficult replacement, and cost. Furthermore, it is of utmost importance to prevent excessive heat generation during device operation, and for this, it is crucial for the inductive link and power management circuitry to maintain very high PTE.

Over the last few decades, several near-field data telemetry methods using the inductive link have been proposed to both power the receiver (Rx) and establish a wireless data link between the Rx and an external transmitter (Tx). Figure 1.2 illustrates a block diagram of a wireless power and data transfer (WPDT) system using the inductive link. Near-field data telemetry methods are advantageous for devices

B. Lee et al., *Energy-management Integrated Circuit Design for Wireless Power and Data Transfer Applications*, Analog Circuits and Signal Processing,
https://doi.org/10.1007/978-3-032-00745-2_1

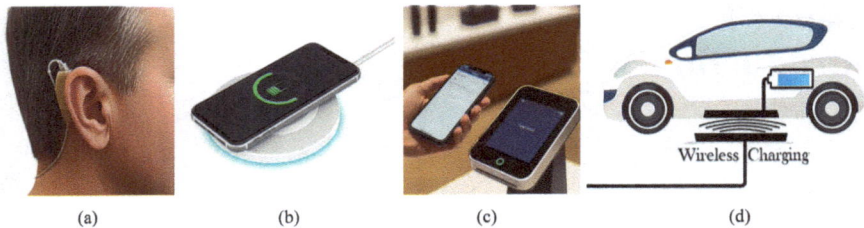

(a)   (b)   (c)   (d)

Fig. 1.1 Various applications of WPT. (**a**) A cochlear implant [1], (**b**) a mobile device charger [2], (**c**) an NFC device [3], and (**d**) wirelessly charging electric vehicles [4]

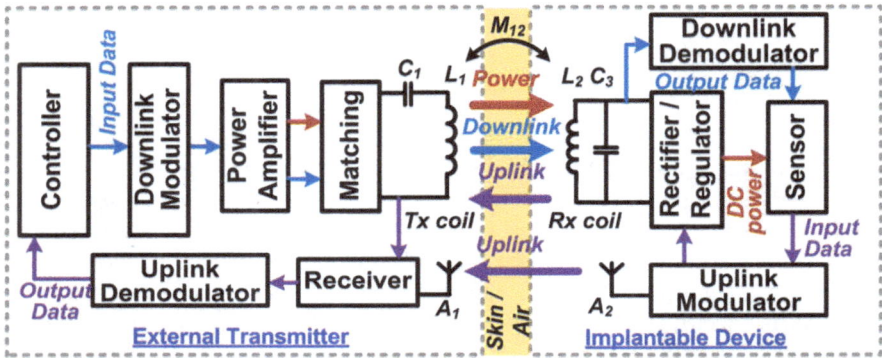

Fig. 1.2 A simplified block diagram of the power and data transfer system through the inductive link [5]

with strict power and area constraints, as they exhibit lower power consumption and simpler circuitry compared to conventional far-field data telemetry methods. These telemetry methods can be categorized based on their characteristics, such as data rate, distance, uplink/downlink, and power consumption. Generally, proper implementation of the telemetry link requires a robust and efficient data link while considering various related factors.

Here are some practical considerations for different types of data telemetry. Data rate and carrier frequency are key related parameters that depend on the application and system architecture, including signal or data processing in the Rx or Tx. While a higher carrier frequency can increase bandwidth, it may also result in more heat dissipation within the device's circuitry at high operation frequencies. Data telemetry distance is generally designed to operate within the centimeter range. The distance needs to be appropriately set based on the coil's geometric elements. Considering the device's achievable power consumption, a data telemetry method requiring an appropriate power level should be adopted. Area is clearly important on the device side, with a strong desire to minimize the number of off-chip components. While some data telemetry methods are composed of simple circuits, applications requiring higher performance may adopt methods that necessitate more complex circuits and additional coils. Reliability or robustness is a crucial factor for stable system operation. In the development of robust data links, the characterization of bit error rate (BER) concerning wireless power conditions or distance variations is of utmost importance.

1.2 Overview

1.2.1 Chapter 2: Basic of Wireless Power Transfer

This chapter provides an understanding of the fundamental principles and circuit modeling of WPT technology. WPT can be classified into near-field techniques utilizing electromagnetic induction and far-field techniques using electromagnetic radiation. Most current wireless charging technologies are based on inductive coupling. This chapter elaborates on the magnetic flux phenomenon in 2-coil inductive links and defines key parameters such as the coupling coefficient (k) and mutual inductance (M). It also covers the equivalent circuit model and efficient design considerations for 2-coil inductive links to optimize PTE and power delivered to load (PDL). Specifically, it emphasizes that the choice between series and parallel configurations for the Rx LC-tank depends on the Rx's quality (Q) factor and the required voltage-current characteristics at the load. Furthermore, multi-coil systems like 3-coil and 4-coil inductive links are introduced, explaining how they can enhance WPT system efficiency and flexibility through improved impedance transformation and higher Q-factors, even under weak coupling conditions. Subsequently, the chapter analyzes the basic principles, design considerations, and common types of DC-AC converters (power amplifiers) essential for generating high-frequency AC current in the Tx coil in WPT systems. Finally, it discusses the role of AC-DC converters (rectifiers) in converting AC signals from the Rx LC-tank into DC power for the load, along with their various configurations.

1.2.2 Chapter 3: Wireless Energy Management Integrated Circuits

This chapter delves into a comprehensive range of rectification architectures and adaptive energy reception techniques, offering insight into both fundamental and state-of-the-art solutions. First, this chapter begins with passive rectifiers, introducing conventional diode-bridge and cross-coupled topologies, and progresses toward enhanced efficiency through threshold voltage-reduction and active diode implementations. Building upon this, we explore active rectifiers that leverage high-speed comparators and offset-controlled feedback for precision switching, significantly improving power conversion efficiency (PCE) at high operating frequencies. The concept of active voltage doublers and multipliers further extends voltage conversion capabilities while addressing startup and timing challenges. To support modern system-on-chip (SoC) applications, we present single- and multi-output resonant regulating rectifiers (R^3), including dual- and three-output designs that efficiently manage power distribution across varying voltage domains. Adaptive rectifier structures, including reconfigurable and optimal-tracking converters, are also covered, demonstrating real-time adjustment to dynamic load and link conditions. Finally, this chapter introduces resonant-mode energy receivers, including time-interleaved and non-residual designs, which enable efficient long-range power delivery by capturing and transferring maximum energy without losses. Together, this chapter provides a foundation and roadmap for designing next-generation AC-DC converters and energy receivers in WPT systems, emphasizing efficiency, adaptability, and integration.

1.2.3 Chapter 4: Wireless Data Telemetry

This chapter focuses on efficient energy management and reliable data telemetry techniques in inductive link-based WPDT systems. Compared to far-field data communication, near-field data telemetry offers lower power consumption and simpler circuit designs, making it suitable for compact wireless applications. This chapter categorizes and explains data telemetry systems based on downlink, uplink, and bidirectional transmission methods. It introduces various modulation techniques such as single-carrier, multi-carrier, pulse-based, and harmonic-based, discussing the characteristics of each. Optimized WPDT systems require a balance of multiple factors, including data transmission rate, communication distance, robustness against link variations, and energy efficiency. Achieving reliable data telemetry in applications with power and size constraints is a significant challenge. To address this, recent WPDT systems focus on integrating a power supply with bidirectional data telemetry through inductive links. These systems efficiently combine power and data transfer to minimize additional components and power consumption while supporting simultaneous downlink and uplink data streams in various wireless applications. Therefore, designers must carefully select the most appropriate data telemetry system, considering the practical limitations of the application and available resources such as power consumption, physical area, data rate, reliability, and communication distance.

1.2.4 Chapter 5: System-Level Wireless Energy Management Techniques

As WPT technologies developed, especially for use in biomedical and implantable devices, system-level energy management became a critical aspect of ensuring secure, efficient, and reliable operation. This chapter introduces advanced techniques that extend beyond basic power conversion, focusing on how energy is intelligently managed, reused, and adapted in real-world applications. First, this chapter presents a secure energy backup receiver designed for cryptographic wireless authentication tags. These tags play a vital role in protecting against counterfeiting in supply chains. A novel energy backup unit (EBU), equipped with non-volatile flip-flops and a clock-controlled voltage doubler, ensures secure data retention even during intentional power disruptions like power-glitch attacks. Next, the optimal reuse energy receiver addresses inefficiencies in conventional wireless back telemetry used in medical implants. Instead of dissipating energy during data transmission, the proposed system temporarily stores the telemetry energy and reuses it for device operation. An adaptive dual-input low-dropout (LDO) regulator intelligently selects between direct and stored energy sources, improving energy efficiency during telemetry and enabling simultaneous power reception and data transmission. Lastly, application-specific WPT solutions are explored through an adaptive power system for deep brain stimulation (DBS). Traditional fixed-voltage designs are replaced with a closed-loop adaptive rectifier, which adjusts its output in real time based on the requirements of stimulation sites. A voltage readout channel enables precise feedback, ensuring that only the necessary power is delivered, thereby improving safety and energy efficiency. Together, these system-level strategies mark a significant step forward in the practical deployment of wireless power for secure, efficient, and intelligent device operation.

References

1. B.S. Wilson, M.F. Dorman, Cochlear implants: A remarkable past and a brilliant future. Hear. Res. **242**(1–2), 3–21 (2008)
2. W.X. Zhong, X. Liu, S.Y.R. Hui, A novel single-layer winding array and receiver coil structure for contactless battery charging systems with free-positioning and localized charging features. IEEE Trans. Ind. Electron. **58**(9), 4136–4144 (2011)
3. Near field communication (NFC) forum. http://www.nfc-forum.org. Accessed 19 Feb 2014
4. C.-S. Wang, O.H. Stielau, G.A. Covic, Design considerations for a contactless electric vehicle battery charger. IEEE Trans. Ind. Electron. **52**(5), 1308–1314 (2005)
5. B. Lee, M. Ghovanloo, An overview of data telemetry in inductively powered implantable biomedical devices design and implementation of devices. IEEE Commun. Mag. **57**(2), 74–80 (2019)

Chapter 2
Basic of Wireless Power Transfer

2.1   Introduction

Since Nikola Tesla's wireless power transfer experiment in Colorado Springs in 1899, wireless power transfer (WPT) technology has steadily advanced and is now utilized in a wide range of applications, including RFID tags, mobile phones, electric vehicle charging, drones, and biomedical implants. WPT technology can be categorized into two main methods based on the mechanism of power transfer: electromagnetic induction using near-field techniques and electromagnetic radiation using far-field techniques.

As shown in Fig. 2.1, the electromagnetic induction method can be further divided into electrodynamic induction, which utilizes magnetic fields, and electrostatic induction, which utilizes electric fields. Among these, most of the wireless charging technologies in use today are based on the inductive coupling method [1]. Recent advancements in WPT technology have enabled its application across a wide range of power levels, from kilowatt-level transmission for electric vehicles (EVs) [2, 3] and kitchen appliances to watt-level transmission for mobile electronics [4, 5], and milliwatt-level transmission for biomedical and wearable devices [6, 7]. While the key focus areas of WPT technology may vary depending on the specific application, this chapter aims to introduce the common theoretical principles and circuit modeling of wireless power transfer using inductive links. Additionally, it explores various techniques to enhance the efficiency of WPT systems.

B. Lee et al., *Energy-management Integrated Circuit Design for Wireless Power and Data Transfer Applications*, Analog Circuits and Signal Processing,
https://doi.org/10.1007/978-3-032-00745-2_2

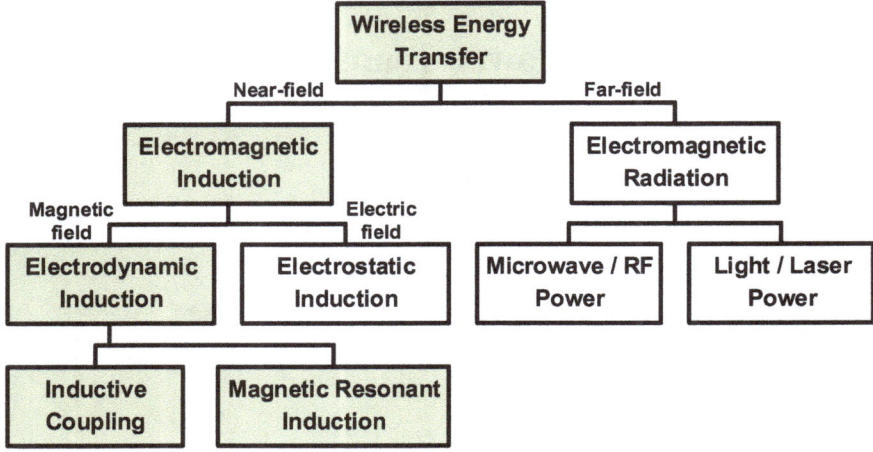

Fig. 2.1 Classification of wireless power transfer methods

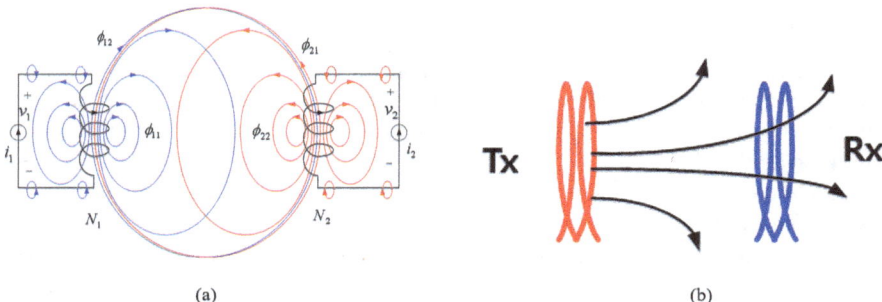

Fig. 2.2 (**a**) Magnetic flux lines in a 2-coil inductive link and (**b**) conceptual diagram of magnetic coupling between Tx and Rx coils

2.2   Near-Field WPT Using Inductive Link

Figure 2.2a conceptually illustrates the magnetic flux in a 2-coil inductive link. When a current i_1 flows through the Tx coil, which has N_1 turns, it generates a magnetic flux. A portion of this flux (Φ_{12}) passes through the Rx coil, which has N_2 turns, while the remaining portion (Φ_{11}) does not pass through the Rx coil and instead loops back to the Tx coil. The portion of magnetic flux (Φ_{12}) that passes through the Rx coil induces a current i_2 in accordance with Faraday's law. The magnetic flux generated by i_2 can similarly pass back through the Tx coil (Φ_{21}) or return to the Rx coil without crossing the Tx coil (Φ_{22}).

As shown in Fig. 2.2b, the coupling coefficient, k, is defined by the proportion of the magnetic flux generated by the Tx coil that passes through the Rx coil. This coefficient is directly determined by the physical geometry of the Tx and Rx coil pair, as shown in Eq. (2.1). The value of $k = 0$ indicates no coupling between the coils, while $k = 1$ represents perfect coupling.

$$k = \left(\frac{1}{\frac{x^2}{r_{Tx}r_{Rx}} + \frac{r_{Tx}}{r_{Rx}}} \right)^2 \tag{2.1}$$

Here, r denotes the radius of the coils, and x represents the distance between the Tx and Rx coils. In most practical cases, the Tx coil is designed to be larger than the Rx coil. It is important to note that k is independent of the number of turns or the material of the coils. Additionally, the magnetic field that does not pass through the Rx coil is not wasted; it loops back without consuming energy. In WPT systems where the coupling coefficient k is low, efficiency loss is caused by energy dissipation in the parasitic resistance of the coil inductors, rather than the loss of the magnetic field itself.

The ability of two coils to magnetically couple and transfer energy is characterized by the mutual inductance (M), which incorporates the physical characteristics of the coupling coefficient k. Mutual inductance, as shown in Eq. (2.2), represents the extent to which the magnetic flux generated by the current in the Tx coil induces a voltage in the Rx coil.

$$M = k\sqrt{L_1 L_2} \tag{2.2}$$

Here, L_1 and L_2 are the self-inductances of the Tx and Rx coils, respectively.

2.3   Circuit Model of 2-Coil Inductive Link

In a typical inductive link, the Tx side is composed of the Tx coil inductance L_{Tx}, coil resistance R_{Tx}, and a resonant capacitor C_{Tx}, as shown in Fig. 2.3. Assuming that the Rx coil inductance L_{Rx} is coupled to the Tx coil with a coupling coefficient k, the

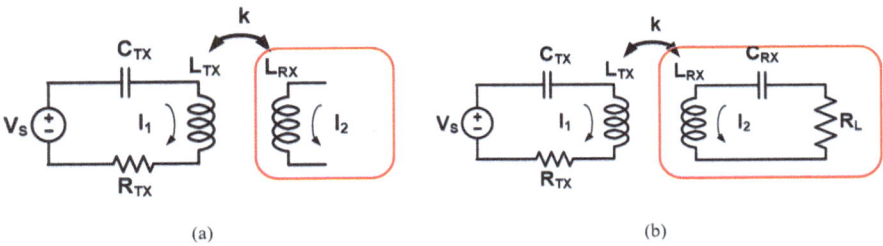

Fig. 2.3 The equivalent circuit model of a 2-coil inductive link for (**a**) Tx side and (**b**) Rx side

equivalent circuit for the Tx side can be derived using Kirchhoff's voltage law (KVL), resulting in the following equation:

$$V_S = I_1 R_{Tx} + \frac{1}{j\omega C_{Tx}} I_1 + j\omega L_{Tx} I_1 + j\omega k \sqrt{L_{Tx} L_{Rx}} I_2 \qquad (2.3)$$

Similarly, for the Rx side, with a load resistance R_L and an LC-tank comprising L_{Rx} and C_{Rx}, KVL can be applied to derive the following equation:

$$0 = j\omega L_{Rx} I_2 + j\omega k \sqrt{L_{Tx} L_{Rx}} I_1 + \frac{1}{j\omega C_{Rx}} I_2 + I_2 R_L \qquad (2.4)$$

Using Eqs. (2.3) and (2.4), the equivalent impedance Z_{out} as seen from the voltage source on the Tx side can be calculated as:

$$Z_{out} = \frac{V_S}{I_1} = R_{Tx} + \frac{1}{j\omega C_{Tx}} + j\omega L_{Tx} + j\omega k \sqrt{L_{Tx} L_{Rx}} \frac{I_2}{I_1} \qquad (2.5)$$

By solving for I_2/I_1 in Eq. (2.5) and substituting it back, the equivalent circuit as seen from the Tx side is modified to include the reflected impedance from the Rx side. This is expressed as:

$$R_{refl} = \frac{\omega^2 k^2 L_{Tx} L_{Rx} / R_{Rx}}{j \frac{\omega L_{Rx}}{R_{Rx}} \left(1 - \frac{\omega^2_{Rx}}{\omega^2}\right) + 1} \qquad (2.6)$$

To maximize the reflected impedance seen from the Tx side, typical inductive link WPT systems ensure that the resonance frequency of the Rx LC-tank (ω_{Rx}/ω_{Rx}) matches the operating frequency of the wireless power carrier (ω). This alignment compensates for the imaginary impedance components, leaving only the real impedance R_{refl}, as shown in:

$$R_{refl} = k^2 \omega L_{Tx} \frac{\omega L_{Rx}}{R_L} \qquad (2.7)$$

assuming that the parasitic resistance of the Rx coil, R_{rp}, is much smaller than R_L.

Figure 2.4 and Eq. (2.7) show that the power delivered to the reflected load, R_{refl}, as seen from the equivalent circuit on the Tx side, is equal to the power delivered to the Rx side. To achieve efficient wireless power transfer, it is crucial to maximize R_{refl}. Given a fixed k, ω, and L_{Tx} in a wireless power transfer scenario, maximizing R_{refl} requires maximizing the Q-factor of the Rx side, defined as:

$$Q = \omega L_{Rx} / R_L \qquad (2.8)$$

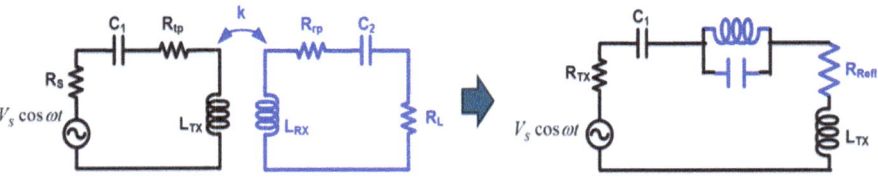

Fig. 2.4   The equivalent circuit of reflected Rx impedance at the Tx side

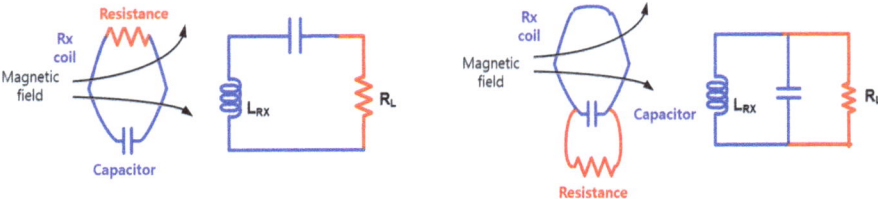

Fig. 2.5   (**a**) Series and (**b**) parallel configurations of Rx LC-tank

To achieve this, it is important to consider the ratio of inductance (L_{Rx}) to resistance (R_L) on the Rx side.

In a typical LC resonant tank, when the resistance R is connected in series within the resonant loop, as shown in Fig. 2.5a, it obstructs the current flow in the resonant tank, leading to reduced power transfer efficiency. Therefore, when R is placed in series in an LC-tank, minimizing the series resistance is advantageous.

Conversely, when the LC-tank is configured in parallel, as shown in Fig. 2.5b, increasing the parallel resistance value brings the circuit closer to an ideal LC-tank. If R is too small, the LC-tank is effectively disrupted. In this case, the quality factor is given as the reciprocal of the series resonance value, expressed as:

$$Q = R_L \, / \, \omega L_{Rx} \qquad (2.9)$$

As a result, maximizing the Q-factor of the Rx side is key to maximizing R_{refl}. To achieve this, the ratio of ωL_{Rx} to R_L must be carefully analyzed. If R_L is relatively small, a series resonance configuration is preferred. On the other hand, if R_L is relatively large, a parallel resonance configuration is more suitable. By optimizing the Rx LC-tank resonance in this way, the reflected resistance R_{refl} seen from the Tx side can be maximized, improving the overall efficiency of the wireless power transfer system.

The choice between series and parallel configurations for the Rx LC-tank is not only determined by the Rx Q-factor but also by the voltage–current characteristics required on the Rx side. As shown in Fig. 2.6a, the equivalent circuit of a series LC-tank on the Rx side can be modeled using the open-circuit (OC) voltage, V_{OC}, to represent the influence from the Tx side. Here, V_{OC} can be expressed in terms of the current flowing through the Tx coil and the mutual inductance. When the Rx

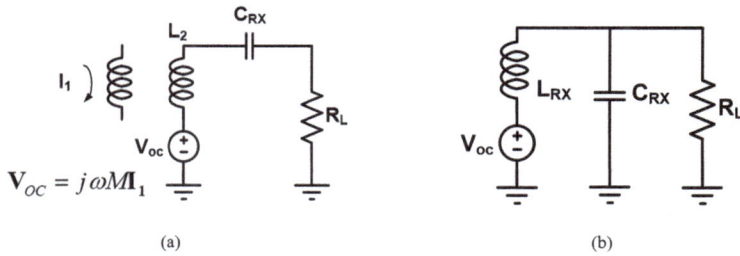

Fig. 2.6 (a) Series LC-tank configuration for constant voltage output and (b) parallel LC-tank configuration for constant current output

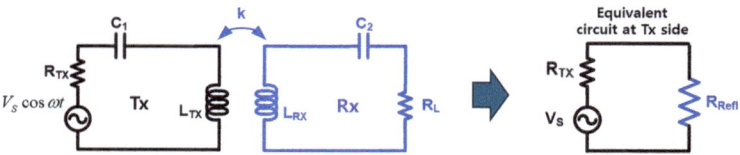

Fig. 2.7 The equivalent circuit of reflected Rx impedance at the Tx side

LC-tank is tuned to the frequency, ω, the V_{OC} is entirely applied to R_L, providing a constant voltage output at the load.

In contrast, if the LC-tank is configured in parallel, a constant current output can be achieved at the load, and the voltage at the actual load can exceed V_{OC} due to the nature of the parallel resonance as shown in Fig. 2.6b. This provides flexibility in tailoring the voltage–current characteristics based on the specific requirements of the Rx system.

2.4 Efficient Optimization in Inductive Link

As shown in Fig. 2.3, if the parasitic resistance of the Rx coil is sufficiently small compared to R_L and can be neglected, the 2-coil inductive link can be represented as shown in Fig. 2.7. If the Tx side LC-tank (L_{Tx}, C_1) is also tuned to resonate at the power carrier frequency, the overall wireless power transfer efficiency will be simply determined by the ratio of the source resistance + parasitic resistance on the Tx side (R_{Tx}) to the reflected resistance (R_{refl}) as follows, which is referred to as the Tx–Rx transfer efficiency:

$$\eta_{\text{Tx–Rx}} = \frac{R_{\text{refl}}}{R_{\text{TX}} + R_{\text{refl}}} = \frac{k^2 Q_{\text{Tx}} Q_{\text{Rx}}}{1 + k^2 Q_{\text{Tx}} Q_{\text{Rx}}} \tag{2.10}$$

Here, the quality factors of the Tx and Rx coils are defined as: $Q_{\text{Tx}} = \omega L_{\text{Tx}}/R_{\text{Tx}}$, $Q_{\text{Rx}} = \omega L_{\text{Rx}}/R_L$.

Equation (2.10) implies that all power delivered to R_{refl} is used by the load R_L in the Rx. This shows that even if the coupling coefficient k between the coils is not high enough, high transfer efficiency can still be achieved by increasing the Q-factor of the coils.

However, in practical scenarios, the parasitic resistance of the Rx coil is often not negligible compared to the load resistance R_L. When this parasitic resistance (R_{PRX}) is included in the equivalent circuit calculations, it becomes necessary to account for the efficiency within the Rx coil itself. The total wireless power transfer efficiency (PTE) is then obtained by multiplying the Tx-Rx transfer efficiency from Eq. (2.10) with the Rx side internal efficiency, which accounts for the load and parasitic resistance:

$$\eta_{PTE} = \frac{R_{refl}}{R_{Tx} + R_{refl}} \frac{R_L}{R_L + R_{PRX}} = \frac{k^2 Q_{TX} Q_{RX}}{1 + k^2 Q_{TX} Q_{RX}} \frac{R_L}{R_L + R_{PRX}} \tag{2.11}$$

In this case, the quality factors of the Tx and Rx coils are redefined as: $Q_{Tx} = \omega L_{Tx}/R_{Tx}$, $Q_{Rx} = \omega L_{Rx}/(R_L + R_{PRX})$.

When considering PTE that includes internal efficiency on the Rx side, as shown in Eq. (2.11), it becomes evident that there exists an optimal load resistance for achieving maximum efficiency. If the parasitic resistance R_{PRX} of the Rx coil is large, the optimal load resistance must increase to account for the receiver's internal efficiency. It is important to note that achieving maximum PTE does not necessarily correspond to the resulting maximum power delivered to the load (PDL).

From the equivalent circuit in Fig. 2.7, delivering power to R_{refl} with maximum efficiency requires maximizing R_{refl}. However, as R_{refl} increases, the actual power delivered to R_{refl} decreases. According to the maximum power transfer theorem in circuit theory, the value of R_{refl} for delivering maximum power is equal to R_{Tx}. Under this condition, the power transfer efficiency to R_{refl} is 50% (since equal power is dissipated in R_{Tx} and R_{refl}.

Therefore, in WPT systems, the PDL can be high, but if the input power at the Tx side is also high, the overall efficiency of the system may decrease. This indicates that maximum PDL does not necessarily mean maximum PTE. In WPT systems, optimizing the load involves determining whether to prioritize maximum PTE or maximum PDL, as shown in Eq. (2.12) and Eq. (2.13), respectively.

$$R_{L,max\,PTE} = R_{PRX} \sqrt{1 + \frac{k^2 \omega L_{Tx} \omega L_{Rx}}{R_{PTX} R_{PRX}}} \tag{2.12}$$

$$R_{L,max\,PDL} = R_{PRX} \left(1 + \frac{k^2 \omega L_{TX} \omega L_{RX}}{R_{PTX} R_{PRX}} \right) \tag{2.13}$$

If the DC-AC converter (RF power amplifier or inverter) circuit is not considered, one might assume that the load resistance R_L can be set for maximum PTE in

Fig. 2.7, and the required power for the load can simply be supplied by increasing the input power from Tx. However, in practical systems where a switching converter is used for DC-AC conversion, the power dissipated in the switches is generally proportional to the parasitic capacitance of the switches (C), the switching frequency (f), and the square of the DC supply voltage (V_{DD}) resulting in $\sim fCV_{DD}^2$.

Thus, setting R_{refl} to its maximum value and simply increasing the power amplifier (PA) V_{DD} to deliver higher power will reduce the DC-AC conversion efficiency, leading to a decrease in overall system efficiency. Furthermore, excessive power increases in the PA require switches with higher voltage ratings, which are often more expensive and may introduce additional unexpected losses in the circuit. Therefore, careful consideration is necessary to avoid such issues and maintain system efficiency.

2.5 Multi-coil Inductive Link

In the previous chapters, we examined how to select appropriate load resistance for maximizing PTE or PDL using circuit modeling of WPT systems. However, in practical scenarios, the load resistance is often dictated by the specific requirements of the Rx application and cannot be arbitrarily adjusted to optimize WPT efficiency. This necessitates the use of impedance transformation techniques to achieve desired efficiency and power delivery levels.

While conventional impedance matching methods using lumped LC networks are widely adopted, they often face limitations due to the intrinsic performance of the LC components, such as their low Q-factor, which impacts overall efficiency. To address these limitations, multi-coil inductive links comprising three or four coils are increasingly utilized [8]. These systems provide inherent benefits, such as enhanced impedance transformation and higher Q-factors, which lead to improved efficiency and flexibility in WPT systems.

The 3-coil inductive link introduces an intermediary resonator coil (L_3) between the Tx coil (L_2) and Rx coil (L_4). The equivalent circuit model and conceptual figure of a 3-coil system are shown in Fig. 2.8. Based on the given circuit diagram of a 3-coil inductive link, the overall power transfer efficiency ($\eta_{3\text{-coil}}$) can be calculated as the product of the efficiencies between each coupled stage of the system:

Fig. 2.8 3-coil inductive link [8]

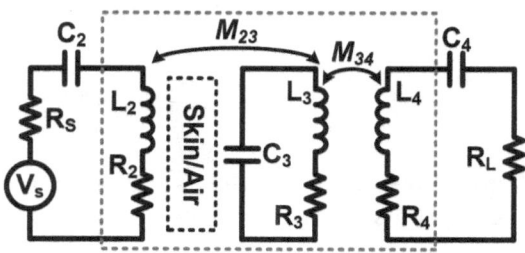

$$\eta_{3-\text{coil}} = \eta_{23} \cdot \eta_{34} \tag{2.14}$$

where η_{23} is the efficiency from the Tx to resonator coil and η_{34} is the efficiency from the resonator coil to the Rx coil (load coil). Therefore, the overall equation from 3-coil inductive link can be derived as shown in Eq. (2.15) similarly from 2-coil inductive link in Fig. 2.7.

$$\eta_{3-\text{coil}} = \frac{\left(k^2_{23}Q_2Q_3\right)\left(k^2_{34}Q_3Q_{4L}\right)}{\left[\left(1+k^2_{23}Q_2Q_3+k^2_{34}Q_3Q_{4L}\right)\left(1+k^2_{34}Q_3Q_{4L}\right)\right]} \cdot \frac{Q_{4L}}{Q_L} \tag{2.15}$$

where $Q_2 = \omega L_2/(Rs + R_2)$ is quality factor of the primary coil, $Q_3 = \omega L_3/R_3$ is quality factor of the resonator coil, $Q_{4L} = \omega L_4/R_L$ and $Q_L = \omega L_4/(R_4 + R_L)$ is quality factor of the load coil, respectively. Compared to the PTE equation of the 2-coil inductive link shown in Eq. (2.11), the 3-coil inductive link introduces an additional degree of freedom (DoF) by incorporating the resonator coil. This additional resonator allows for the adjustment of k_{34}, the coupling coefficient between the resonator coil and the Rx coil. For example, in the conventional 2-coil inductive link, when the Tx coil, Rx coil, and load resistance are predetermined, achieving maximum PTE requires the use of a matching circuit to transform the load impedance. However, in the 3-coil inductive link, simply adjusting the coupling (k_{34}) between the resonator coil and the Rx coil can transform the fixed load resistance into the optimal reflected resistance (R_{ref}) required for maximum PTE, eliminating the need for complex matching circuits.

The 4-coil inductive link extends the 3-coil configuration by adding an additional intermediary coil (L_2) on the Tx side (near the power source). This setup provides even greater flexibility in impedance matching and further isolates the source and load. The equivalent circuit and coil configuration of 4-coil system are shown in Fig. 2.9. The overall power transfer efficiency ($\eta_{4-\text{coil}}$) can be derived similarly with the equation of 3-coil inductive link:

$$\eta_{4-\text{coil}} = \eta_{12} \cdot \eta_{23} \cdot \eta_{34} \tag{2.16}$$

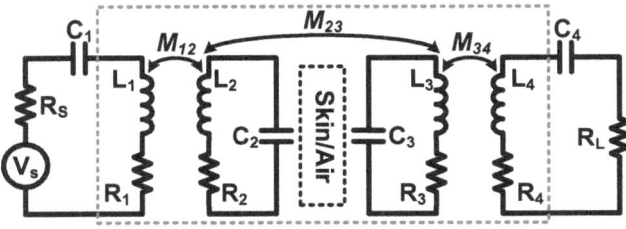

Fig. 2.9 4-coil inductive link [8]

where η_{12} is the efficiency from the Tx coil to the Tx resonator coil, η_{23} is the efficiency from the Tx resonator coil to the Rx resonator coil, η_{23} is the efficiency from the Rx resonator coil to the load coil. Therefore, the overall PTE equation of a 4-coil inductive link can be derived as:

$$\eta_{4-coil} = \frac{\left(k^2_{12}Q_1Q_2\right)\left(k^2_{23}Q_2Q_3\right)\left(k^2_{34}Q_3Q_{4L}\right)}{\left[\left(1+k^2_{12}Q_1Q_2\right)(1+k^2_{34}Q_3Q_{4L})+k^2_{23}Q_2Q_3\right]\left[1+k^2_{23}Q_2Q_3+k^2_{34}Q_3Q_{4L}\right]} \cdot \frac{Q_{4L}}{Q_L} \quad (2.17)$$

where $Q_1 = \omega L_1/(Rs + R_1)$, $Q_2 = \omega L_2/R_2$, $Q_3 = \omega L_3/R_3$, $Q_{4L} = \omega L_4/R_L$, and $Q_L = \omega L_4/(R_L + R_4)$. As shown in Eq. (2.17), in a 4-coil inductive link, increasing k_{12} generally enhances the PTE while sacrificing the PDL.

The selection of 2-coil, 3-coil, or 4-coil inductive links typically depends on factors such as the coupling strength between the Tx and Rx coils, the required PDL, and the system's robustness to coupling variations. As described in [8], 2-coil links are generally advantageous under strong coupling conditions, while 4-coil links, with their ability to perform source and load impedance transformation, become more favorable as the coupling weakens. For instance, 2-coil links are predominantly used in applications requiring large PDLs (in the W range or higher), such as wireless charging for mobile devices and smartphones, where the coupling between Tx and Rx is relatively strong. On the other hand, in applications involving weak coupling, such as small-scale biomedical implants operating in the mW range or below, 3-coil or 4-coil links are often preferred due to their ability to transform the load impedance effectively. However, implementing 3-coil or 4-coil links requires the addition of intermediary coils, which increases the complexity and size of the overall WPT system. Therefore, practical feasibility must be considered when choosing these configurations. Table 2.1 summarizes the characteristics and suitability of different coil configurations based on system requirements and operating conditions.

2.6 DC-AC Converter

In WPT systems, the DC-AC converter plays a critical role in generating high-frequency alternating currents required for power transmission through the Tx coil. This conversion is typically achieved using power amplifiers (PAs) or inverters, depending on the system's application, operating frequency, and efficiency requirements. This section discusses the fundamental principles, design considerations, and common types of DC-AC converters used in WPT systems.

Since the WPT system requires high-frequency AC signals to drive the Tx coil, the DC-AC converter takes the DC supply voltage (from batteries or other sources) and converts it into a high-frequency AC waveform suitable for wireless power transmission. The choice of DC-AC converter impacts the system efficiency,

Table 2.1 Comparison of 2-coil, 3-coil, and 4-coil inductive links for WPT

	2-coil link	3-coil link	4-coil link
Strong coupling (k)	High efficiency but limited flexibility	Good efficiency with added flexibility	Possibly good efficiency with complex control
Weak coupling (k)	Significant efficiency loss	Moderate performance improvement	Best performance due to additional impedance matching
Large PDL (small R_s)	High-power delivery but requires precise tuning	Handles large PDL well with reasonable efficiency	Can be optimized for large PDL, but more losses due to complexity
Small PDL (large R_s)	Efficiency drops significantly	Moderate performance	Maintains higher efficiency due to impedance control
Coupling variations and small R_s	Highly sensitive to variations	Improved robustness	Best robustness against coupling variations
Coupling variations and large R_s	Severe performance losses	Moderate improvements	Best mitigation of coupling variations
Additional components and area	Small	Moderate	High

available frequency range, output power capacity, complexity, and cost. Power amplifiers (PAs) are generally categorized into two types: analog signal amplifiers and digital switching amplifiers. Analog amplifiers, such as Class-A, Class-B, Class-AB, and Class-C, amplify signals in a continuous manner, preserving their shape. On the other hand, digital switching amplifiers, including Class-D, Class-E, and Class-F, convert DC to AC through high-speed on/off switching operations.

In a Class-A amplifier, the transistor conducts throughout the entire signal cycle (360° conduction). This results in excellent linearity and minimal distortion, making it suitable for high-fidelity signal amplification. However, it suffers from significant drawbacks, including low efficiency (~25–30%) due to constant conduction and high-power dissipation, leading to excessive heat generation. As a result, Class-A amplifiers are rarely used in WPT systems where efficiency is a critical factor.

The Class-B amplifier improves upon Class-A by having the transistor conduct for only half of the signal cycle (180° conduction). Typically, two transistors are used to handle the positive and negative halves of the signal. This design increases efficiency to around 50% but introduces significant crossover distortion at the point where one transistor turns off and the other turns on. While Class-B amplifiers are suitable for some audio applications, they are rarely employed in WPT systems due to their moderate efficiency and distortion issues.

The Class-AB amplifier combines the advantages of Class-A and Class-B designs by allowing the two transistors to slightly overlap in conduction, reducing crossover distortion. This results in better linearity than Class-B and higher efficiency (~50–60%) than Class-A. While Class-AB amplifiers strike a good balance between efficiency and linearity, they are not commonly used in WPT systems, as their efficiency is still not high enough for most power transfer applications.

The Class-C amplifier takes efficiency even further by having the transistor conduct for less than half of the signal cycle (<180° conduction). This design achieves very high efficiency (~75–85%) but at the expense of poor linearity. A resonant circuit is typically used to restore the full signal waveform, making Class-C amplifiers more suitable for high-frequency applications such as RF systems. They are sometimes used in WPT systems operating in high-frequency ranges, but are not ideal for low-power or low-frequency scenarios.

In contrast to analog amplifiers, Class-D amplifiers use digital switching techniques to achieve very high efficiency (~90%). The transistors in a Class-D amplifier switch fully on and off, minimizing power dissipation. This design generates square waveforms, which are typically filtered to produce the desired AC signal. Class-D amplifiers are widely used in consumer WPT systems, such as wireless charging pads, due to their simplicity, cost-effectiveness, and high efficiency. However, they have moderate linearity compared to analog amplifiers.

The Class-E amplifier is a specialized switching amplifier that employs Zero Voltage Switching (ZVS) to minimize switching losses. By shaping the transistor's voltage and current waveforms, Class-E amplifiers achieve extremely high efficiency (~90–95%) and reduced component stress. However, they are sensitive to parameter variations and require precise tuning to maintain optimal performance. Class-E amplifiers are commonly used in low-power, high-efficiency WPT systems, such as biomedical implants and wearable devices.

Finally, the Class-F amplifier uses harmonic tuning networks to shape the output waveform, minimizing overlap between voltage and current waveforms for maximum efficiency. Class-F amplifiers are capable of delivering very high efficiency and high power, but come with increased design complexity. They are typically used in high-frequency and high-power WPT systems where precise harmonic control is required.

Each amplifier class offers unique trade-offs in terms of efficiency, linearity, and complexity, making them suitable for different WPT applications based on system requirements, as shown in Table 2.2. For example, low-power biomedical devices often favor Class-E amplifiers, while consumer charging systems rely on Class-D amplifiers for their simplicity and efficiency.

2.7 AC-DC Converter

In WPT systems, circuit modeling, as shown in Fig. 2.7, typically involves optimizing reflected resistance through the load resistance to maximize either PTE or PDL. However, in practical scenarios, the Rx LC-tank is not directly connected to a load resistance. Instead, the AC signal is converted to DC through an AC-DC conversion circuit before being delivered to the load.

Table 2.2 Comparison of power amplifier classes

	Conduction angle	Efficiency	Linearity	Applications
Class-A	360°	Low (~25–30%)	Excellent	High-fidelity audio systems (not WPT)
Class-B	180°	Moderate (~50%)	Poor (crossover distortion)	Audio amplification (not WPT)
Class-AB	180°–360°	Moderate (~50%)	Good	Audio systems (rarely WPT)
Class-C	<180°	High (~75–85%)	Poor (requires tuning)	RF WPT systems, high-frequency applications
Class-D	Switching	Very High (~90%)	Moderate	Consumer WPT, wireless charging pads
Class-E	Switching (ZVS)	Very High (90–95%)	Moderate	Biomedical implants, low-power WPT
Class-F	Switching (Harmonic)	Very High	Moderate	High-power, high-frequency WPT systems

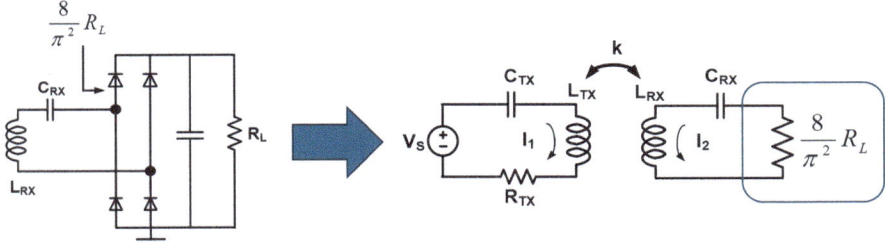

Fig. 2.10 Equivalent load impedance using conventional full-wave rectifier

Common AC-DC conversion circuits include half-wave rectifiers and full-wave rectifiers, which utilize single or multiple diodes. To minimize the voltage drop caused by the diodes, active rectifiers using MOSFET switches are sometimes employed. Additionally, voltage doublers are often used to boost the output voltage in wearable or biomedical applications. The detailed circuit structures and operating principles of these various AC-DC converters will be discussed in later chapters of this book.

In this chapter, we focus on analyzing the changes in load impedance for AC load when conventional full-wave rectifier configuration, as shown in Fig. 2.10. This analysis will provide insights into how load impedance affects the overall system performance in practical WPT applications.

If the Rx LC-tank is configured in a series structure as shown in Fig. 2.10, and a square wave voltage $V_R(t)$ is applied to the Rx LC-tank while a full-wave rectifier is connected as shown in Fig. 2.11, the square wave voltage $V_R(t)$ can be represented using a Fourier series:

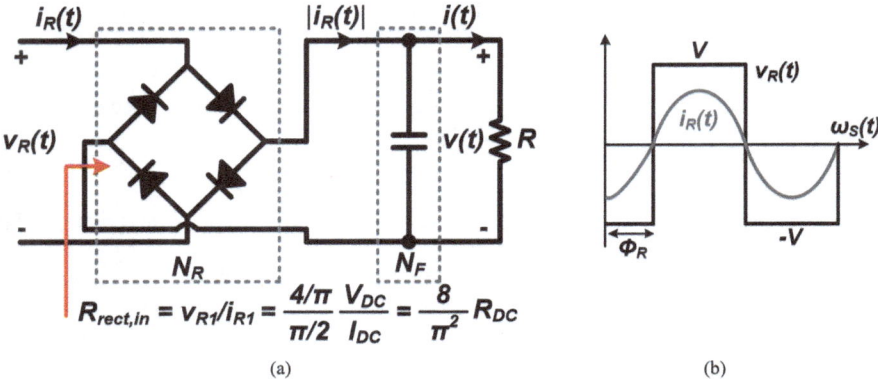

$$R_{rect,in} = v_{R1}/i_{R1} = \frac{4/\pi}{\pi/2}\frac{V_{DC}}{I_{DC}} = \frac{8}{\pi^2}R_{DC}$$

(a)   (b)

Fig. 2.11 (**a**) Driving voltage and current waveforms in full-wave rectifier and (**b**) its waveforms

$$V_R\left(t\right) = \frac{4V_{DC}}{\pi} \cdot \sum_{n=1,3,5...} \frac{1}{n}\sin\left(n\omega t\right) \qquad (2.18)$$

Here, V_{DC} is the rectifier output voltage, and n denotes the harmonic order. Assuming the output filter capacitor is sufficiently large, and the ripple is negligible, $V_R(t)$ becomes a square wave with zero crossings that are in phase with the tank output current $i_R(t)$. Since the Rx LC-tank operates only at the resonant frequency ω, the higher-order harmonics are filtered out, and the remaining voltage across the Rx LC-tank, which is the fundamental input voltage of the rectifier ($V_{R,1st}$), can be expressed as:

$$V_{R,1st} = \frac{4}{\pi}V_{DC} \qquad (2.19)$$

The charge balance of the output capacitor implies that the DC load current is equal to the average rectified tank output current:

$$I_{DC} = \frac{2}{\pi}I_{R,1st} \qquad (2.20)$$

Thus, the fundamental current $I_{R,1st}$ is related to the DC load current as follows:

$$I_{R,1st} = \frac{\pi}{2}I_{DC} \qquad (2.21)$$

Here, $I_{R,1st}$ is the fundamental harmonic component of I_{RI}. Therefore, the equivalent input impedance of the rectifier $R_{rect,in}$ can be defined as:

$$R_{\text{rect,in(series)}} = \frac{V_{\text{R,1st}}}{I_{\text{R,1st}}} = \frac{8}{\pi^2} R_{\text{DC}} \qquad (2.22)$$

Here, R_{DC} represents the DC resistance of the rectifier.

In the case where the LC-tank is configured in a parallel connection, the rectifier is driven by a sinusoidal voltage, and the input voltage and current operate inversely compared to the series connection. Therefore, by following a similar sequence to determine the voltage and current relationship as in the series connection, it can be easily understood that $R_{\text{rect,in}}$ can be calculated as the reciprocal of the value in Eq. (2.22), as shown below.

$$R_{\text{rect,in(parallel)}} = \frac{\pi^2}{8} R_{\text{DC}} \qquad (2.23)$$

2.8   Conclusions

This chapter has provided a foundational understanding of WPT technology, with a particular focus on systems utilizing inductive links. Beginning with a historical overview, the chapter established the fundamental principles of near-field WPT, detailing the magnetic flux phenomenon between a Tx and Rx coil and defining key parameters such as the coupling coefficient and mutual inductance. The efficient design of WPT systems based on inductive links requires a comprehensive understanding of these fundamental principles, including coil coupling, impedance matching, power amplifier selection, and load considerations. The insights provided herein serve as a crucial basis for developing robust and highly efficient wireless power solutions across diverse applications.

References

1. A. Laha, A. Kalathy, M. Pahlevani, P. Jain, A comprehensive review on wireless power transfer systems for charging portable electronics. Eng **4**, 1023–1057 (2023)
2. A. Zakerian, S. Vaez-Zadeh, A. Babaki, A dynamic WPT system with high efficiency and high power factor for electric vehicles. IEEE Trans. Power Electron. **35**(7), 6732–6740 (2020)
3. P. Machura, V. De Santis, Q. Li, Driving range of electric vehicles charged by wireless power transfer. IEEE Trans. Veh. Technol. **69**(6), 5968–5982 (2020)
4. H. Jung, B. Lee, Wireless power and bidirectional data transfer system for IoT and mobile devices. IEEE Trans. Ind. Electron. **69**(11), 11832–11836 (2022)
5. Y.-J. Park, Next-generation wireless charging systems for mobile devices. Energies **15**, 3119 (2022)
6. Z. Li, J. Lee, J. Lim, B. Lee, Efficient distributed wireless power transfer system for multiple wearable sensors through textile coil array. Sensors **23**, 2810 (2023)

7. H.-S. Lee, K. Eom, H.-M. Lee, A single-input multi-output resonant regulating rectifier generating three outputs in a half cycle for wirelessly powered biomedical devices. IEEE Trans.
   Circuits Syst. I: Regul. Pap. (2025). https://doi.org/10.1109/TCSI.2025.3531211
8. M. Kiani, U.-M. Jow, M. Ghovanloo, Design and optimization of a 3-coil inductive link for efficient wireless power transmission. IEEE Trans. Biomed. Circuits Syst. 5(6), 579–591 (2011)

Chapter 3
Wireless Energy Management Integrated Circuits

3.1 Active AC-DC Converter

3.1.1 Passive Rectifier

When the L_2C_2 resonant tank receives AC voltages, four diodes are traditionally used to convert the AC voltage into DC voltage for powering applications [1–7]. In CMOS processes, these diodes can be implemented using diode-connected PMOS or NMOS transistors, as illustrated in Fig. 3.1a [1]. The diode-connected PMOS transistors (P_{1P} and P_{2P}) handle the connection between the L_2C_2 tank and the rectifier output voltage (V_{OUT}), while the diode-connected NMOS transistors (N_{1P} and N_{2P}) connect the tank to V_{SS}. However, this configuration results in a significant voltage drop across V_{OUT}, equivalent to four threshold voltages ($2V_{THN} + 2V_{THP}$), which reduces the power conversion efficiency (PCE), defined as the ratio of output DC power to input AC power. To address this issue, the diode-connected NMOS transistors can be replaced with a cross-coupled NMOS pair (N_{3P} and N_{4P}), as shown in Fig. 3.1b [2]. This design allows the NMOS pair to turn on and off in response to the AC input voltages, reducing the dropout to $2V_{THP}$ and improving PCE. Still, the voltage dropout caused by the diode-connected PMOS transistors (P_{3P} and P_{4P}) continues to impact PCE negatively.

To further reduce the dropout in V_{OUT}, the diode-connected PMOS transistors can be replaced with V_{TH}-reduced diodes as depicted in Fig. 3.2 [3]. During startup, V_{OUT} initially reaches $V_{IN} - V_{TH,D2}$ via D_2. Simultaneously, the capacitor C_C is charged to $V_{OUT} - V_{TH,D3}$ through D_3, and then V_C becomes $V_{IN} - V_{TH,M1}$. Eventually, V_{OUT} is defined as $V_C + V_{TH,D2}$, which simplifies to $V_{IN} - V_{TH,M1} + V_{TH,D3}$. By incorporating V_{TH}-reduced diodes into the NMOS cross-coupled rectifier, the dropout voltage can be reduced to $2(V_{TH,D3} - V_{TH,M1})$. However, some dropout voltages remain, and the inclusion of the auxiliary capacitor increases the overall chip size.

© The Author(s), under exclusive license to Springer Nature
Switzerland AG 2025
B. Lee et al., *Energy-management Integrated Circuit Design for Wireless Power and Data Transfer Applications*, Analog Circuits and Signal Processing,
https://doi.org/10.1007/978-3-032-00745-2_3

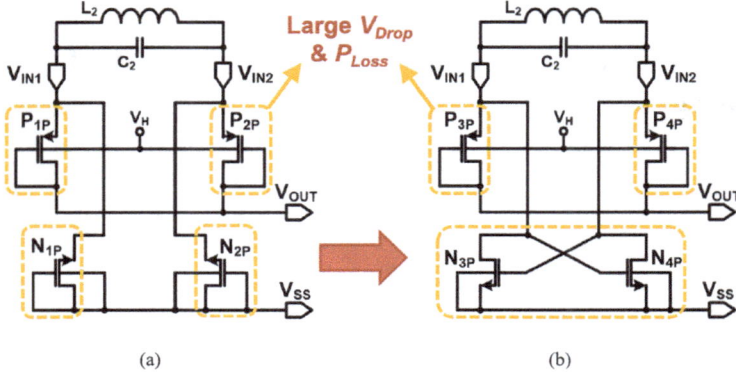

Fig. 3.1 Schematics of (**a**) the diode-connected bridge rectifier and (**b**) the NMOS cross-coupled rectifier

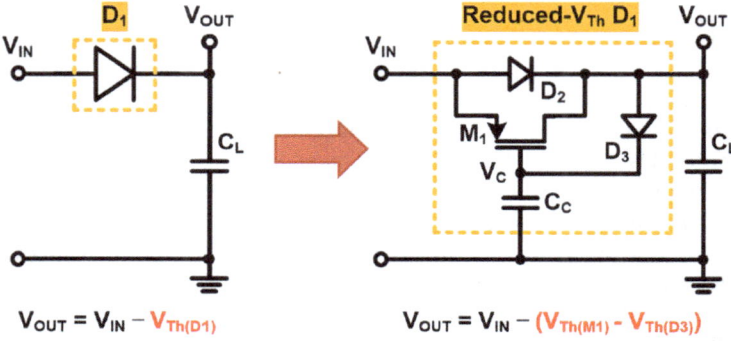

Fig. 3.2 Schematic of the V_{TH}-reduced diode

To eliminate the need for an auxiliary capacitor, the passive diode can be replaced with an active diode consisting of a switch controlled by a comparator, as shown in Fig. 3.3 [4]. The comparator activates the switch only when the input AC voltages (V_{IN1} and V_{IN2}) exceed V_{OUT} by comparing V_{IN} and V_{OUT}, reducing the voltage drop to <100 mV. This minimal dropout significantly enhances PCE. However, if the CG comparator fails to precisely detect the point where the AC voltage equals V_{OUT}, delayed turn-on or turn-off timings can introduce idle periods and allow back current to flow from the output node to the L_2C_2 tank, ultimately lowering PCE.

3.1.2 Active Rectifier

The active full-wave active rectifier employs a pair of high-speed comparators (CMP$_1$ and CMP$_2$) to control the main rectifying elements (P_1 and P_2), as shown in Fig. 3.4 [5]. The input voltage of the rectifier, $V_{IN} = V_{IN1} - V_{IN2}$, follows a sinusoidal

Fig. 3.3 Conceptual
diagram of the active diode

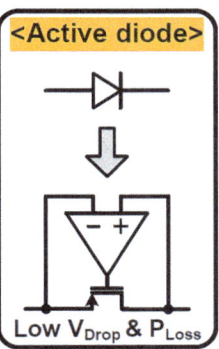

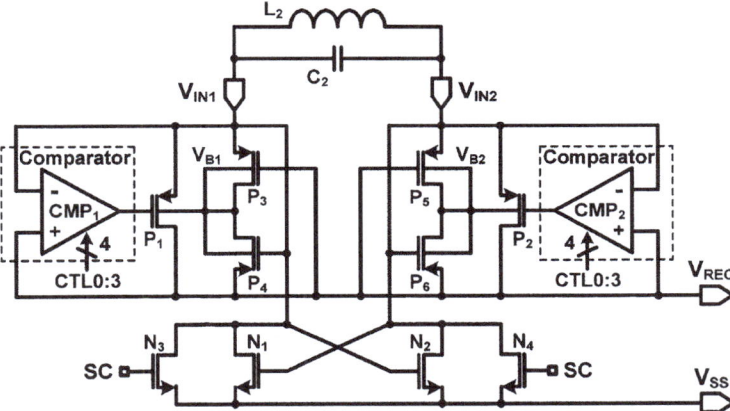

Fig. 3.4 Schematic diagram of active rectifier including offset-controlled high-speed comparators [5]

waveform, causing P_1 and P_2 to alternate their operation based on the polarity and amplitude of V_{IN}. When V_{IN} exceeds the NMOS threshold voltage (V_{THN}) but is less than the rectified voltage (V_{REC}), the positive feedback loop of the cross-coupled NMOS pair (N_1 and N_2) connects V_{IN2} to V_{SS} through N_2, turning off N_1. At this point, CMP_2 outputs a high signal, indicating that $V_{REC} > V_{SS}$, and P_2 is turned off. P_1 remains off as long as V_{IN} remains below V_{REC}. Once V_{IN} exceeds V_{REC}, CMP_1 outputs a low signal, turning P_1 on and allowing current to flow from V_{IN1} to V_{REC}, charging the rectifier's resistive/capacitive load ($R_L C_L$). During the next half cycle, when V_{IN} falls below V_{THN}, V_{IN1} is connected to V_{SS} through N_1, turning off N_2, while both P_1 and P_2 remain off until V_{IN} rises above V_{REC} again. At this point, CMP_2 activates, turning P_2 on, and current flows from V_{IN2} to V_{REC} to charge the load once more. To prevent latch-up and substrate leakage issues between P_1 and P_2, the potentials at their separate N-well body terminals (V_{B1} and V_{B2}) must always be the highest on-chip. This is achieved through dynamic body bias control, utilizing auxiliary PMOS transistors (P_3 to P_6), as described in [6]. This technique ensures that

V_{B1} and V_{B2} are automatically connected to the highest potential, whether from the input voltages (V_{IN1} and V_{IN2}) or the rectified output voltage (V_{REC}), thus enhancing efficiency and safety.

To drive the large rectifying PMOS transistors at the high operating frequency of 13.56 MHz, high-speed comparators with low power consumption and strong driving capability are essential. Typically, the speed of a comparator is constrained by its propagation delay (T_P), which determines how quickly the output responds to input changes. In this rectifier application, propagation delay negatively impacts power conversion efficiency (PCE). Due to the high-to-low transition delay (T_{PHL}), the comparators turn P_1 and P_2 on too late, reducing the input power that could have been transferred to the load during this period. Additionally, due to the low-to-high transition delay (T_{PLH}), the comparators are slow in turning off P_1 and P_2, allowing current to briefly flow from the load capacitor (C_L) back to the secondary coil when $V_{IN} < V_{REC}$.

Since it is impossible to eliminate T_P entirely, an offset-control function in the high-speed comparators is implemented in this rectifier to mitigate these limitations. Figure 3.5 shows the schematic of the high-speed comparator with two offset-control functions, labeled offset F and offset R. Excluding the offset-control blocks and cross-coupled inverters (CS inverters), the comparator operates as a basic common-gate comparator with start-up capability. The input voltages, V_{REC} and V_{IN1}, are applied to the sources of the input transistors P_7 and P_8, respectively. When V_{IN1} exceeds V_{REC}, the current through P_8 surpasses that of P_7, causing the gate voltage of the output inverter (V_A) to rise rapidly, which lowers V_{OUT} and turns on P_1. The offset F and offset R blocks are implemented using current sources (P_{13}–P_{14} and P_{10}–P_{11}), multiplexers (MUXs), and control switches (P_{15} and P_{12}). These blocks inject offset currents into the comparator inputs alternately, adjusting the switching timing. For example, when V_{OUT} is high, P_{15} turns on, and an offset current flows into

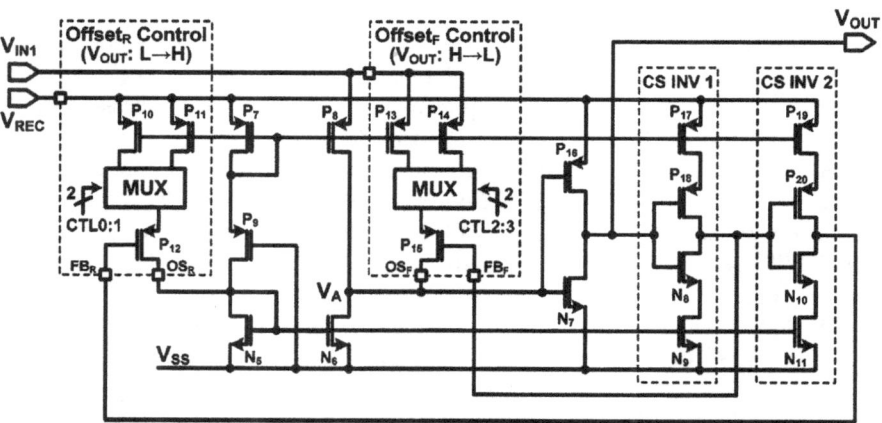

Fig. 3.5 Schematic diagram of active rectifier including offset-controlled high-speed comparators [5]

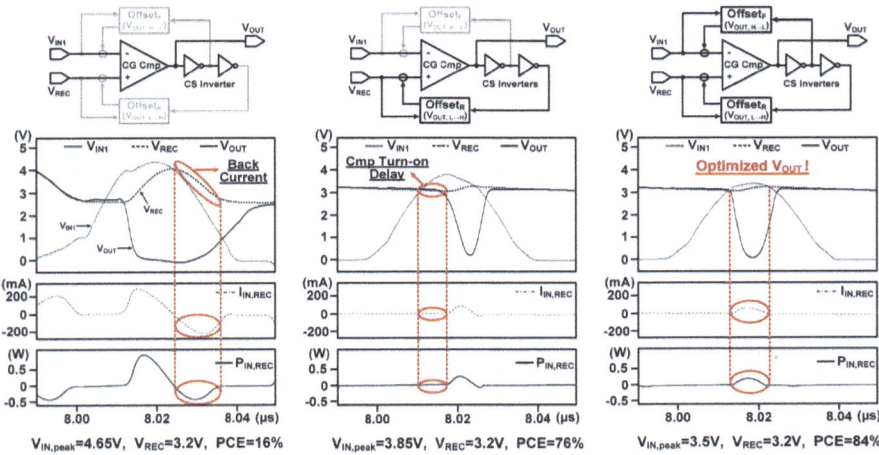

Fig. 3.6 Schematic diagram of active rectifier including offset-controlled high-speed comparators [5]

the positive input branch of the comparator (V_{REC}) via the offset F block, causing V_A to increase. As a result, V_{OUT} begins to fall earlier, even before V_{IN1} exceeds V_{REC}. The magnitude of the offset current is programmable using 2-bit off-chip control signals (CTL0:1 for offset F and CTL2:3 for offset R), allowing the rectifier timing to be fine-tuned in response to process variations.

Simulation results illustrating the relationship between power conversion efficiency (PCE) and the offset-control functions are shown in Fig. 3.6. To better understand the effects of these functions, the rectifier's input/output voltages, input current, and input power waveforms were overlaid, while adjusting the V_{IN} amplitude to maintain a constant output voltage (V_{OUT}) of 3.2 V for a load resistance (R_L) of 500 Ω. With no comparator offset-control function, the turn-off delay leads to significant back current, which severely degrades the PCE. However, with the use of the offset R function effectively prevents this back current, resulting in a noticeable improvement in PCE. Despite this enhancement, the input power to the rectifier is still compromised due to the turn-on delay (T_{PHL}) of the comparators. This indicates that further optimization of the rectifier's PCE and voltage conversion ratio (VCR) is possible by addressing both the turn-on and turn-off delays. By employing both the offset F and offset R functions, which compensate for T_{PHL} and T_{PLH}, respectively, the rectifier's performance is significantly improved. Thus, with both offset-control functions active, the V_{OUT} transitions occur at optimal times, maximizing the PCE.

Figure 3.7 presents the measured power conversion efficiency (PCE) as a function of load resistance (R_L) with V_{REC} set to 3.12 V, a load capacitance (C_L) of 10 μF, and an operating frequency (f_C) of 13.56 MHz. As R_L increases, output power of the rectifier decreases for the same V_{REC}. Consequently, the internal power dissipation due to switch losses and power consumption of comparator becomes more

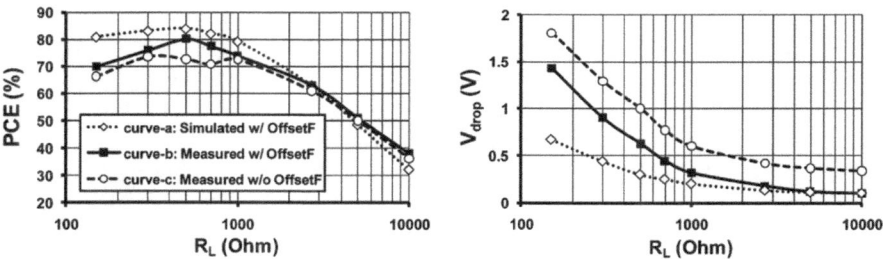

Fig. 3.7 Measured PCE and V_{drop} versus R_L with $V_{REC} = 3.12$ V, $C_L = 10$ μF, and $f_C = 13.56$ MHz [5]

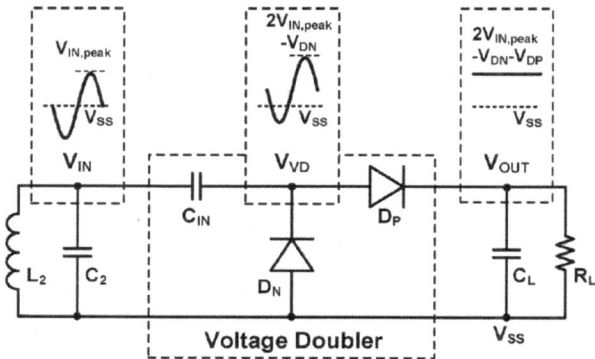

Fig. 3.8 Schematic diagram of the passive voltage doubler using diodes [7]

significant, which reduces the overall PCE. Figure 3.7 also shows both measured and simulated voltage drop (V_{drop}) as a function of R_L, demonstrating that V_{drop} decreases as R_L increases. This is because a larger R_L requires a smaller rectifier current (I_{REC}), resulting in a smaller voltage drop across the rectifying transistors.

3.1.3 Active Voltage Doubler

Figure 3.8 illustrates the topology of a conventional passive voltage doubler, which can be implemented using diodes. The circuit consists of an input capacitor (C_{IN}) and two diodes, D_N and D_P, each with corresponding forward voltage drops, V_{DN} and V_{DP} [7]. The rectified output voltage, V_{OUT}, is low-pass filtered by the output capacitor (C_L) and supplies the load resistor (R_L). The sinusoidal input voltage (V_{IN}) generated across the secondary resonance circuit (L_2C_2) has a peak amplitude ($V_{IN,peak}$) which is determined by the inductive link parameters and the power amplifier (PA) output. When V_{IN} falls below $-V_{DN}$, V_{VD} is connected to V_{SS} through D_N, charging C_{IN} to $V_{IN,peak} - V_{DN}$, with V_{VD} as the positive node. As V_{IN} rises above $-V_{IN,peak}$, D_N turns off, and V_{VD} increases to follow $V_{IN,peak} - V_{DN} + V_{IN}$. When V_{VD} exceeds

$V_{OUT} + V_{DP}$, D_P turns on, allowing current to flow from V_{IN} to V_{OUT}, charging the $R_L C_L$ load. During this phase, the charge stored in C_{IN} decreases as it is delivered to the load, but C_{IN} is recharged to $V_{IN,peak} - V_{DN}$ in the next cycle. Due to the voltage drop across D_P, the maximum output voltage (V_{OUT}) is limited to $2V_{IN,peak} - V_{DN} - V_{DP}$. The total voltage drop across the doubler, V_{drop}, can be expressed as $V_{drop} = 2V_{IN,peak} - V_{OUT} = V_{DN} + V_{DP}$, indicating that the forward voltage drops of D_N and D_P directly influence the output voltage and, consequently, the power conversion efficiency (PCE). Therefore, replacing these diodes with fast MOS switches that have lower on-resistance and leakage would effectively reduce V_{drop} and improve the PCE.

To drive N_1 and P_1 at high frequencies, such as 13.56 MHz, the comparators are equipped with adjustable internal offset-control functions (CTL0:3) to minimize the effects of comparator delay. Additionally, the separated N-well body terminal of P_1 must be connected to the highest potential on the chip to avoid latch-up and substrate leakage issues. Thus, as shown in Fig. 3.9, a dynamic body biasing technique, using auxiliary transistors P_3 and P_4, which automatically connect V_{BODY} to the highest potential between V_{VD} and V_{OUT} was employed. Since the comparators are powered by V_{OUT}, which initially starts at 0 V, it is essential for the active voltage doubler to have startup capability. The startup block generates a complementary pair of start-up enable signals, SU and SU_B, based on the V_{OUT} level, to control the startup switches N_2 and P_2 as well as the comparators. When V_{OUT} is too low to operate the comparators, the startup circuit sets SU to high and SU_B to low, turning on N_2 and P_2, respectively, and disabling the comparators. In this state, both N_1 and P_1 are

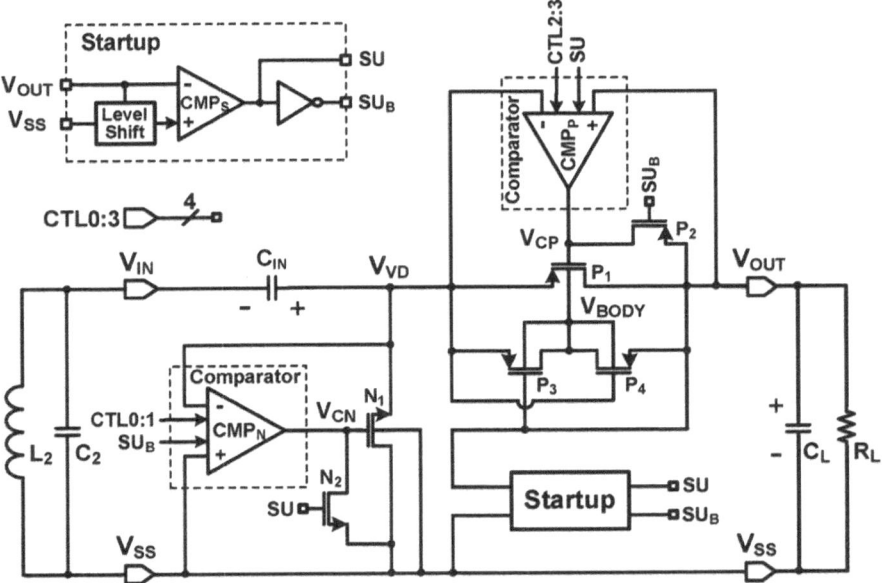

Fig. 3.9 Schematic diagram of the proposed active voltage doubler with offset-controlled comparators [7]

diode-connected, forming a passive voltage doubler that begins charging V_{OUT}, independent of the status of the comparators. Once V_{OUT} reaches a level sufficient to power the comparators, SU and SU_B toggle, turning off N_2 and P_2 and enabling the comparators to activate the active voltage doubler.

Figure 3.10 presents the schematic of two symmetrical high-speed comparators, CMP_N and CMP_P, each featuring three built-in offset-control functions. In CMP_N, transistors P_7–P_8, N_3–N_4, and P_{15}–N_7 form a common-gate comparator, where the input terminals at the sources of N_3 and N_4 are connected to V_{SS} and V_{VD}, respectively. A biasing branch, composed of P_6 and resistor R_1, is mirrored onto P_7 and P_8, ensuring that the comparator requires a minimum supply voltage of $V_{Th(P6)}$ to begin operation. The gate of the diode-connected N_3 is coupled with N_4, meaning the currents through N_3 and N_4 depend on their respective source voltages, V_{SS} and V_{VD}. When V_{VD} is less than V_{SS}, the current through N_4 exceeds that of N_3, P_7, and P_8, causing V_1 (the input to the P_{15}–N_7 inverter) to drop rapidly. This results in a high comparator output voltage (V_{CN}) which turns on N_1.

Although common-gate comparators are generally regarded as high speed due to their low input impedance and simple architecture, their speed was insufficient to drive large capacitive loads (N_1 and P_1) at 13.56 MHz. To address this, Offset-1_N and Offset-2_N circuits are integrated into CMP_N (with their counterparts in CMP_P) to compensate for turn-on and turn-off delays, respectively. The Offset-1_N block uses an N_5 current source controlled by N_6, which pulls additional offset current from output branch of CMP_N, causing V_1 to drop earlier when activated by V_{OS1N} = high. Offset-2_N, a constant offset mechanism, is implemented by size mismatching P_8 and P_7, with the larger W/L ratio of P_8 pushing additional offset current into the output branch to increase V_1 sooner. The offset-control signal V_{OS1N} is generated by an offset-control block, consisting of a current-starved inverter (P_{16}–P_{17}–N_8) and other logic gates. When V_{VD} exceeds V_{SS}, V_{CN} becomes low, and V_{OS1N} goes high. This activates N_6, turning on N_5 to pull offset current in parallel with N_4 at a higher level than the current pushed by Offset-2_N through P_8. Consequently, V_{CN} starts to rise earlier, turning on N_1 slightly before V_{VD} drops below V_{SS}, compensating for the comparator's turn-on delay. Once V_{CN} reaches a high state, the Offset-1_N block turns off, and the offset current through P_8 dominates. As a result, V_{CN} starts to decrease earlier, turning off N_1 slightly before V_{VD} surpasses V_{SS}, compensating for the comparator's turn-off delay. V_{OS1N} goes high after a delay set by the current-starved inverter, which is shorter than one carrier cycle period. Since V_{OS1N} switches to high

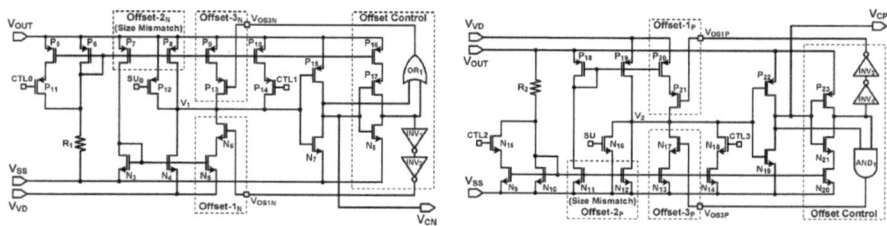

Fig. 3.10 Schematic diagram showing three offset-control functions in CMP_N and CMP_P [7]

when V_{VD} is significantly greater than V_{SS}, it avoids any feedback-induced fluctuation or instability issues.

To further clarify the impact of offset-control functions on power conversion efficiency (PCE), Fig. 3.11 compares simulation results showing the input/output voltages of the voltage doubler (V_{VD} and V_{OUT}), the comparator output voltages (V_{CN} and V_{CP}), and input power waveforms, with and without the offset controls enabled. In these simulations, an AC voltage of $V_{IN,peak} = 2$ V at a frequency of 13.56 MHz was applied to the input, while an $R_L C_L$ load of 1 kΩ and 2 nF was connected to the output of the active voltage doubler. Without the offset-control functions, comparator turn-on delays cause V_{CN} and V_{CP} to activate transistors N_1 and P_1 too late, resulting in power conduction delays through the pass transistors from the $L_2 C_2$ tank to the load when V_{VD} is less than V_{SS} or greater than V_{OUT}. Additionally, turn-off delays lead to V_{CN} and V_{CP} deactivating N_1 and P_1 too late, causing back currents to flow from C_{IN} to V_{SS} and from the output load back to the $L_2 C_2$ tank. These inefficiencies significantly reduce the output voltage ($V_{OUT} = 2.4$ V) and PCE ($\eta = 28\%$). When the offset-control functions are enabled, the waveforms show a substantial reduction in conduction delays and back currents. The Offset-1 and Offset-2 functions correct for the turn-on and turn-off delays, respectively, ensuring that V_{CN} and V_{CP} switch the pass transistors at the optimal times, which maximizes PCE. With these controls in place, the active voltage doubler achieves a much higher output voltage ($V_{OUT} = 3.43$ V) and PCE ($\eta = 80\%$) under the same input voltage and load conditions. Additionally, the Offset-3 function ensures that V_{CN} and V_{CP} remain at V_{SS} and V_{OUT}, respectively, after their conduction periods to reliably turn off the pass transistors, preventing any spurious variations in V_{VD}.

3.1.4 Active Voltage Multiplier

For resonant frequencies ranging from 1 MHz to 40.68 MHz, achieving high PCE in an inductively powered wireless system typically requires an active diode with the offset-control techniques, as detailed in Sects. 3.1.1, 3.1.2, and 3.1.3. However,

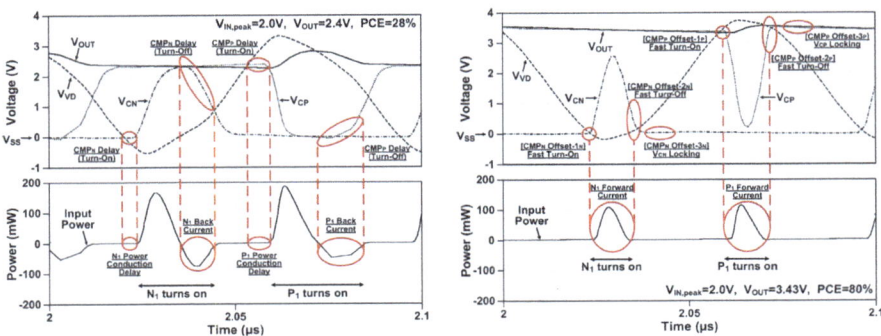

Fig. 3.11 Simulation waveforms of the active voltage doubler without and with offset-control functions [7]

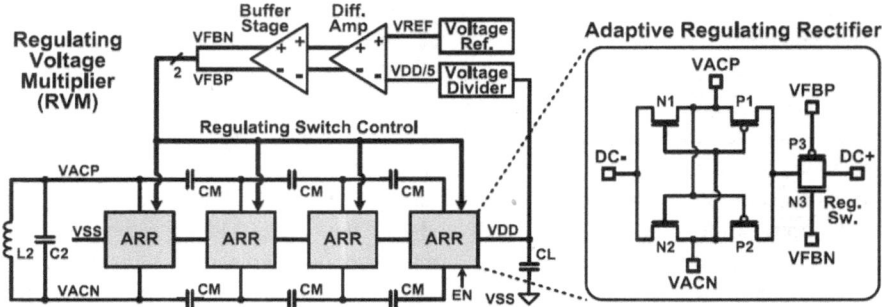

Fig. 3.12 Schematic of the regulating voltage multiplier and adaptive regulating rectifier [8]

at higher frequencies, such as 433 MHz, the offset control demands a significantly higher bias current to accurately detect the point where the AC voltage matches the rectifier output voltage. In such cases, passive diodes become a more efficient alternative to active diodes. For high-frequency passive rectifiers, multi-stage cross-coupled differential rectifiers are commonly employed [8].

To regulate the rectifier output (V_{DD}) to meet the target voltage, an adaptive regulating rectifier with the power regulating switches (N_3 and P_3) is proposed as shown in Fig. 3.12. Unlike the open-loop rectifier described in [9], where the DC output fluctuates with changes in AC input amplitude, the proposed ARR utilizes N_3 and P_3 to maintain a stable DC output level using negative feedback signals (V_{FBN} and V_{FBP}). Then, V_{DD} is divided and compared to a reference voltage through a differential amplifier and buffer. V_{FBN} and V_{FBP} dynamically adjust the dropout voltages across the regulating switches, ensuring that V_{DD} is consistently regulated to the target voltage.

3.2 Adaptive Energy Receiver

3.2.1 Reconfigurable AC-DC Converter

The adaptive reconfigurable voltage doubler (VD)/rectifier (REC) is a versatile circuit that can dynamically switch between functioning as a voltage doubler or a rectifier based on the sensed output voltage, enabling more reliable inductive power transmission over longer distances [10]. Figure 3.13 illustrates the conceptual design of this VD/REC, which integrates two separate AC-DC converters—a rectifier and a voltage doubler—into a unified structure. The operating mode (either rectifier or voltage doubler) is selected by an external mode selection signal. The circuit consists of a full-wave rectifier and a voltage doubler with active diodes. The rectifier requires two diodes, D_1 and D_2, along with a cross-coupled NMOS pair, N_1 and N_2. Depending on the amplitudes of V_{INP} and V_{INN}, either the $D_1 - N_2$ path or

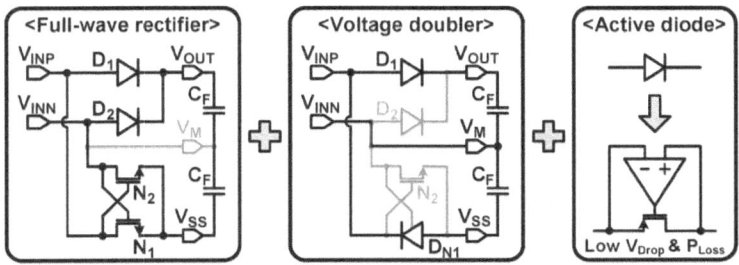

Fig. 3.13 Conceptual diagram of the active VD/REC converter using active diodes [10]

$D_2 - N_1$ path is activated to transfer input power to the output filtering capacitors $C_F/2$. The voltage doubler, on the other hand, requires only two diodes, D_1 and D_{N1}, and charges one C_F per half cycle depending on the polarity and amplitude of V_{IN} ($= V_{INP} - V_{INN}$). This configuration almost doubles the output voltage compared to the peak of V_{IN}. To support both modes, D_1 is shared between the two structures, while D_2 and N_2 are selectively enabled depending on the operating mode. N_1 functions as part of the cross-coupled pair in rectifier mode but is reconfigured as NMOS diode D_{N1} in voltage doubler mode. Additionally, V_{INN} and V_M are shorted through a switch (N_3) when in voltage doubler mode. The VD/REC employs active diodes (D_1, D_2, and D_{N1}), where rectifying pass transistors are controlled by fast comparators, allowing them to operate as switches in the deep triode region with low dropout voltages. This results in lower power dissipation compared to passive diodes, enhancing power conversion efficiency (PCE) in both operating modes.

Figure 3.14 presents a simplified schematic of the VD/REC, which consists of PMOS and NMOS active diodes (D_1, D_2, and D_{N1}) controlled by pass transistors P_1, P_2, and N_1. These transistors are driven by high-speed comparators (CMP$_{P1}$, CMP$_{P2}$, and CMP$_{N1}$) to reduce AC-DC dropout voltage and minimize power loss. The mode signals EN and EN_B, generated by the mode control circuit, configure the VD/REC to operate either in rectification or voltage doubling mode. In rectification mode (EN = 0, EN$_B$ = 1), the gates of N_1 and N_2 are connected to V_{INN} and V_{INP}, respectively, forming a cross-coupled NMOS pair with positive feedback. In this mode, CMP$_{N1}$ is deactivated, and both PMOS active diodes are engaged, alternating every half cycle to transfer power to the load. For example, when V_{INP} exceeds V_{INN}, N_2 turns on, and N_1 turns off. When V_{INP} exceeds V_{OUT}, CMP$_{P1}$ outputs a low signal, turning P_1 on with a low dropout voltage. The input current flows from V_{INP} through P_1, charging the filtering capacitor ($C_F/2$) and returning to V_{INN} via N_2. In voltage doubling mode (EN = 1, EN$_B$ = 0), P_2 and N_2 are turned off, and CMP$_{P2}$ is deactivated. Only P_1 and N_1 remain active, while N_3 connects V_{INN} to V_M, charging the filtering capacitors (C_F) one per half cycle, effectively doubling V_{OUT} relative to V_{SS}. Because the comparators are powered by V_{OUT}, which starts at 0 V, the VD/REC includes a self-start-up feature. The start-up circuit monitors V_{OUT} and forces SU = EN = EN$_B$ = 0 when V_{OUT} is too low. During start-up, N_1 and N_2 are cross-coupled through multiplexers (MUXs), and P_1 and P_2 are connected as diodes

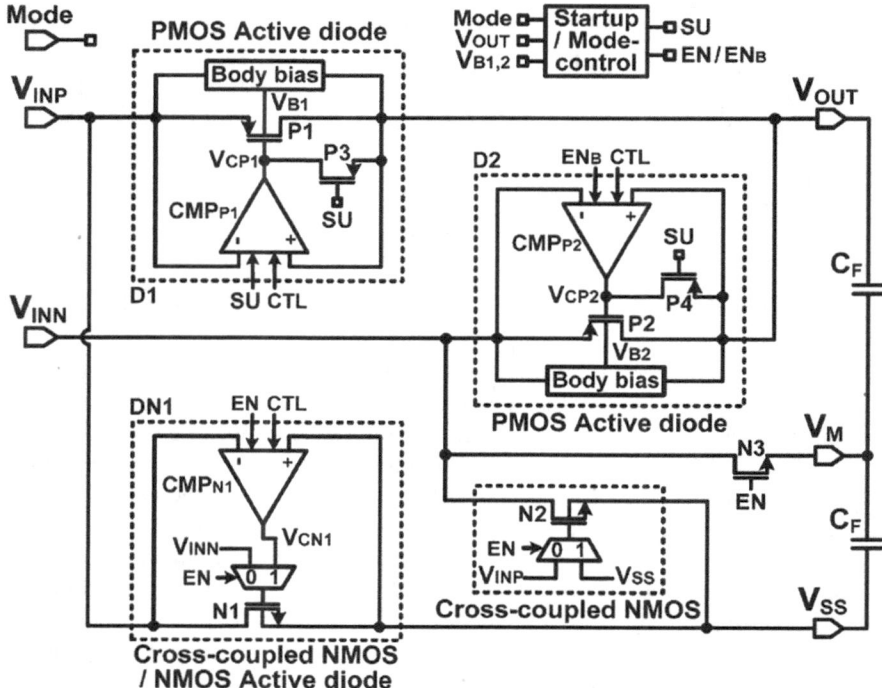

Fig. 3.14 Schematic diagram of the adaptive reconfigurable VD/REC employing active diodes [10]

through P_3 and P_4, respectively, forming a passive rectifier that charges V_{OUT} until it reaches a sufficient level for normal operation. Once V_{OUT} is high enough, SU toggles, allowing the VD/REC to function as designed. Additionally, the PMOS body terminals V_{B1} and V_{B2} are always connected to the highest potential among V_{INP}, V_{INN}, and V_{OUT} via dedicated body bias circuits.

To verify the advantages of using the VD/REC over a standard rectifier (REC), measurements were taken of $V_{IN,peak}$ and V_{OUT} while varying the relative distance (*d*) and orientation (*θ*) between two coupled coils, L_1 and L_2. Figure 3.15 illustrates the measured $V_{IN,peak}$ and V_{OUT} as functions of *d* and *θ*, showing how the VD/REC extends the inductive power transmission range compared to using only a REC. The hysteresis window of the off-chip comparator, set to 2.6–3.7 V, is indicated on the graphs with horizontal dashed lines. In the distance sweep (d-sweep) test, the VD/REC operates in rectifier mode when the coils are close together. As the distance increases, V_{OUT} decreases, and when *d* exceeds 5.5 cm, the VD/REC switches to voltage doubler mode, boosting V_{OUT} by 0.8 V (a 30.8% increase). This allows the VD/REC to maintain a sufficient V_{OUT} (>2.5 V) for coil separations up to 8 cm, compared to the REC, which fails beyond 6 cm, representing a 33% improvement. Similarly, in the angular sweep test, the VD/REC enhances the system's tolerance to coil misalignment. It extends the operational range from

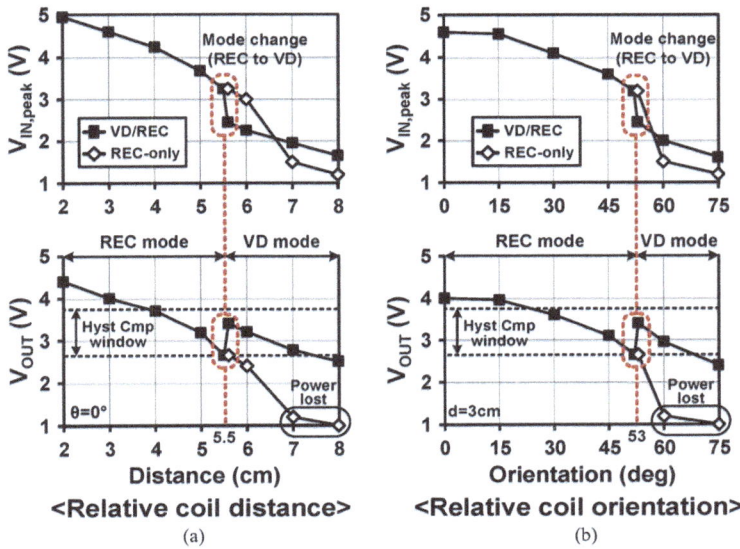

Fig. 3.15 Measured input and output voltages while sweeping (**a**) the coil relative distance and (**b**) orientation [10]

$\theta = 53°$ (REC only) to $\theta = 75°$ (VD/REC) at $d = 3$ cm, reflecting a 41.5% improvement in performance.

3.2.2   Optimal-Tracking AC-DC Converter

Section 3.1.2 explained the full-wave rectifier with the fixed offset control in PMOS power pass transistors and with NMOS cross-coupled pair [11, 12]. Since the mobility of the NMOS is typically better than PMOS ($\times 2.5$ @ 250 nm process), the rectifier with the fixed offset control in NMOS power pass transistor and with PMOS cross-coupled structure as shown in Fig. 3.16 achieves higher PCE than the other one, with the same silicon size. Even though this structure can achieve higher PCE, but still higher PCE could be achieved by eliminating unwanted turn-on/off delays due to fixed offset control. These unwanted turn-of/off delays vary with various circumstances such as input power, output load, and resonant frequency, thus the adaptive offset-control technique is required to achieve high not only PCE, but also VCR.

Figure 3.17 shows the adaptive offset control with analog feedback consisting of two sample and hold (S/H) blocks, two operational transconductance amplifiers (OTA) controlling the offset current (I_{OF}) depending on the error voltage (V_{ER}) [13]. However, the additional power dissipation flowing through two OTAs results in lower PCE. Also, the analog feedback using two OTAs results in limited

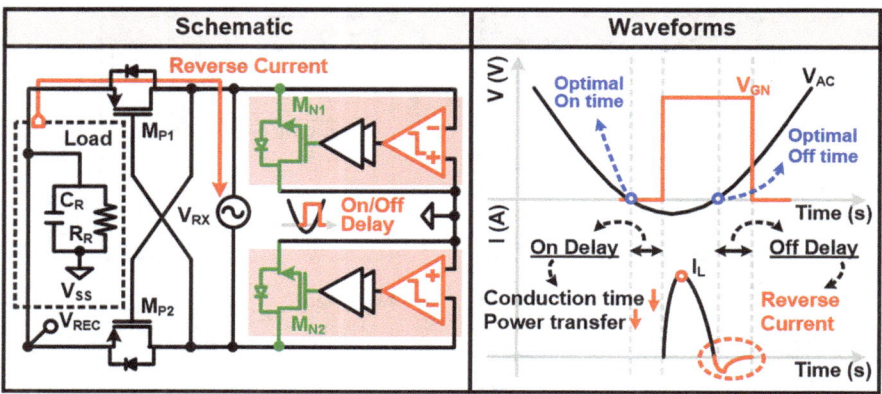

Fig. 3.16 The full-wave rectifier with the fixed offset control in NMOS power pass transistor [13]

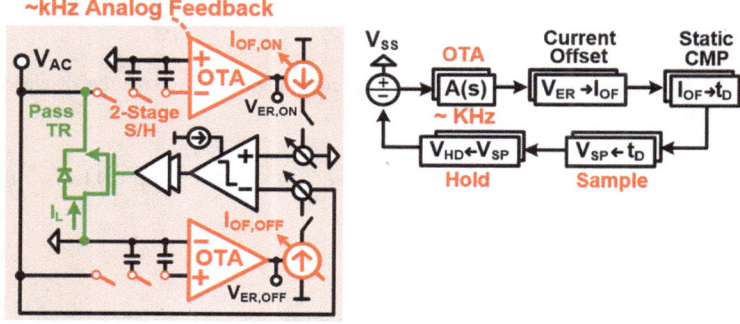

Fig. 3.17 Adaptive offset control with analog feedback [13]

compensation speed under kHz and cannot respond rapidly to on/off error. Even though the use of SAR-ADC or 1.5-bit ADC with current regulators has been replaced with speed-limited OTAs, analog components still restrict loop bandwidth to kHz ranges, limiting ADC speed and sampling rates. This slow response can cause delays in adjusting to changing load and input conditions, reducing efficiency and regulation accuracy. Additionally, the limited dynamic range of OTA requires extra hold stages to handle input variations, affecting precision. For wireless charging of IMDs, where input and load conditions change frequently, fast tracking of optimal offset injections through frequent monitoring and rapid feedback speed is essential to maintain performance and avoid delays.

Figure 3.18 shows the adaptive offset control with digital feedback substituting the analog components of two S/H blocks and two OTAs with two dynamic latched comparators (DLC) and digital controller. The proposed digital feedback-based delay control (DFDC) utilizes digitally controlled delay lines (DCDL) to efficiently generate adaptive on/off transitions for the SR-latch, eliminating the need for

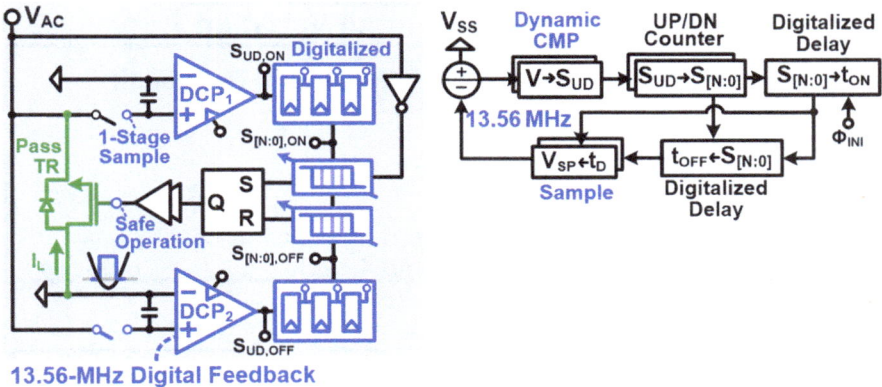

Fig. 3.18 The adaptive offset control with digital feedback using DFDC [13]

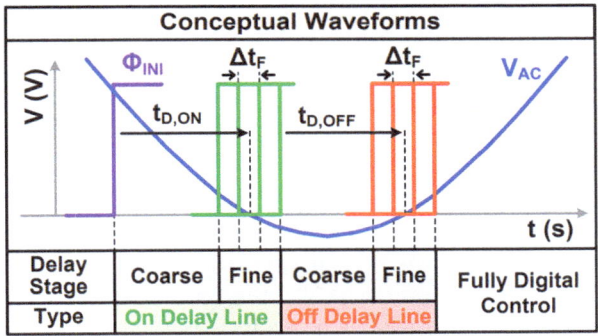

Fig. 3.19 Conceptual waveforms of the coarse and fine delay lines for switch driving [13]

high-power static comparators. The DCDL adjusts the initial clock (Φ_{INI}) to achieve optimal timing based on the digital input data (S[N:0]). This fully digital controller achieves no static loss and furthermore, with real-time power-saving mode, higher PCE could be achieved. Moreover, the high-speed optimal on/off timing tracking due to digital feedback can stabilize the output voltage in a fast loop response.

The coarse and fine delay lines (CFDL) adjust on delay ($t_{D,ON}$) and off delay ($t_{D,OFF}$) based on digital delay data, using coarse delays for rapid changes during startup or varying conditions and fine delays for precise timing with minimal error as shown in Fig. 3.19. Once optimal timing is achieved and the system stabilizes, DFDC activates a power-saving mode by disabling the input clock in non-critical circuits, reducing power loss and enhancing efficiency. This power-saving mode is possible since the delay information is digitally stored, allowing continued optimal operation even during sleep mode.

Figure 3.20a shows the DFDC's input transient operation in digital feedback mode ($S_{MD} = 0$), tested with a 1 kΩ load and input voltage |V_{AC}| varying from 2 V to

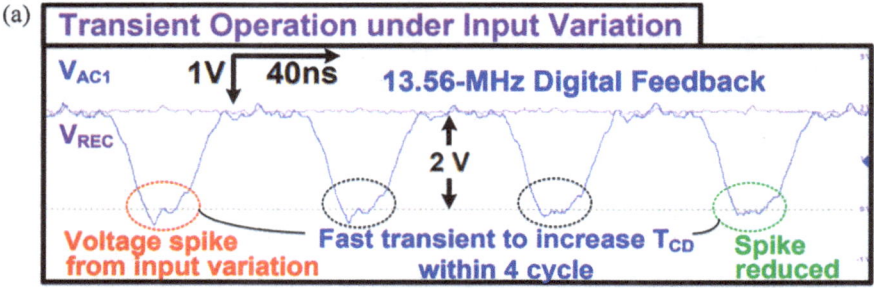

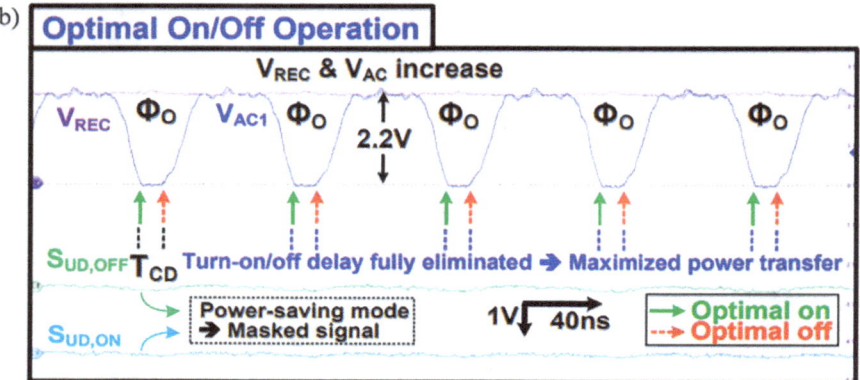

Fig. 3.20 (**a**) The adaptive offset control with digital feedback using DFDC and (**b**) power-saving mode [13]

2.2 V. As coil distances decrease, input power increases, but timing deviations cause voltage spikes due to insufficient conduction time. DFDC minimizes these spikes by adjusting delays using 13.56-MHz digital feedback, stabilizing within four cycles. In steady-state operation ($S_{MD} = 1$), as shown in Fig. 3.20b, the mode controller detects the optimal cycle ($N_{OPT} = 6$), and the dual-path delay selector fine-tunes the on/off points for optimal driving. Once the delay is locked, $S_{UD,ON}$ and $S_{UD,OFF}$ are masked, ensuring continuous operation with precise timing, maximizing power transfer and eliminating voltage spikes.

Figure 3.21a, b illustrate the experimental results for PCE and VCR, respectively. The highest PCE observed was 93.5%, occurring when the $|V_{AC}|$ reached 2.6 V, R_L was set to 0.5 kΩ, and the delivered output power was 12.52 mW. Across a wide load range from 0.33 to 2.2 kΩ, the PCE consistently stayed above 88.3%, highlighting the robustness of the system. This stable performance, with only a 4.3% variation in efficiency, was enabled by dynamic adjustment of the switching control based on output power levels. For the VCR, which reflects how closely the rectified voltage (V_{REC}) follows the input $|V_{AC}|$, the maximum recorded value was 96.3% at $|V_{AC}| = 2.1$ V and $R_L = 2.2$ kΩ. Due to the low-loss characteristics of the DFDC rectification, VCR remained above 94.9% throughout the measured range, with fluctuations limited to just 1.2%.

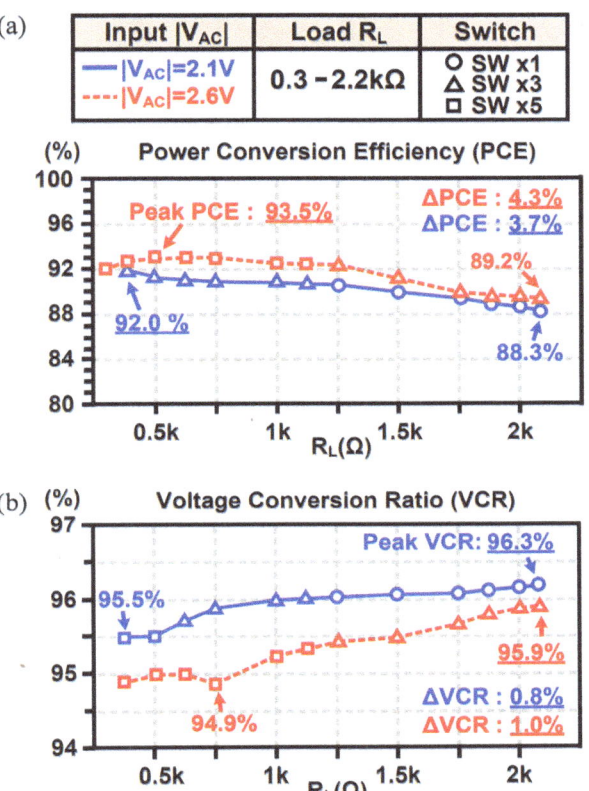

Fig. 3.21 (**a**) Measured PCE and (**b**) measured VCR of the proposed active rectifier [13]

3.3 Multi-Output Regulating Rectifier

3.3.1 Regulating Rectifier

The two-stage AC-DC regulator, which combines an AC-DC rectifier and a DC-DC converter to convert an AC waveform into the desired DC voltage, is a common approach [14]. However, this design has drawbacks, including reduced efficiency due to the two-step conversion process and increased size caused by the large output capacitor in the AC-DC rectifier. To address these issues, the two-stage structure can be streamlined into a more compact and efficient single-stage design [15] (Fig. 3.22).

Most AC-DC rectifiers utilize a CG comparator to identify the moment when V_{IN}(AC) equals V_{REC}. During the turn-on phase, when V_{IN}(AC) exceeds V_{REC}, energy is transferred from the L2C2 tank to the load, leaving V_{REC} undefined as illustrated in Fig. 3.23a. With this method, V_{REC} varies with changes in input amplitude and cannot be adjusted internally. The proposed 1-stage AC-DC regulator with the

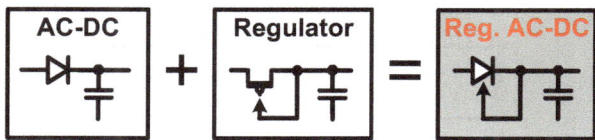

Fig. 3.22 Conceptual diagram of 1-stage AC-DC regulator

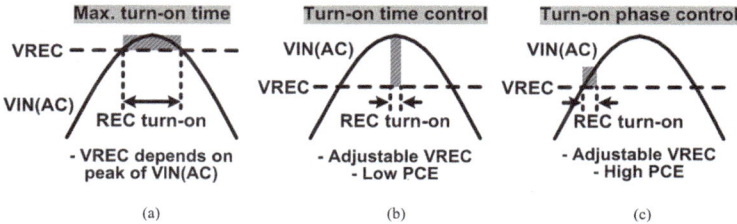

Fig. 3.23 Comparison between (**a**) conventional AC-DC rectifier and 1-stage AC-DC regulator (**b**) with turn-on time control and (**c**) with the turn-on phase control [15]

turn-on time control shows that V_{REC} can be regulated by modifying the turn-on timing near the peak of $V_{IN}(AC)$ as shown in Fig. 3.23b. Shortening the turn-on duration reduces the forward current, which in turn lowers V_{REC}. However, this approach causes a significant voltage drop between $V_{IN}(AC)$ and V_{REC} during the turn-on phase, leading to high-power dissipation in the rectifying transistors and reduced power conversion efficiency (PCE). To achieve adjustable V_{REC} while preserving high PCE, the rectifier turn-on phase is optimized, as depicted in Fig. 3.23c. In this approach, the rectifier activates when $V_{IN}(AC)$ exceeds V_{REC}, similar to conventional designs, but the turn-off timing is adjusted to control the forward current. This method allows fine-tuning of V_{REC}, while minimizing the voltage drop between $V_{IN}(AC)$ and V_{REC} during operation, thereby enhancing PCE.

The turn-off timing in the 1-stage AC-DC regulator can be controlled using a pulse-width modulation (PWM) technique by comparing V_{REC} with the target voltage. The phase control comparator (CMP_1), depicted in Fig. 3.24, integrates components P_4, P_5, N_6, N_7, P_8, and N_8 to create a common-gate comparator that compares V_{REC} and V_{INP}. A current source, P_7, adds current when V_{O1} is high and P_6 is active, accelerating the rise of V_1 and enabling P_1 to transition more quickly. The phase control feedback mechanism utilizes inverter chains and a current-starved inverter (INV_6) paired with N_{10}. The current bias for INV_6 is regulated by AMP_1, which compares $V_{REC}/3$ with V_{REF} to produce the required delay. INV_6 output is further delayed, influencing the turn-off transistor P_3, allowing the rectifier to adaptively shut off even before V_{INP} drops below V_{REC}, ensuring the desired V_{REC} is achieved.

The measured waveforms in Fig. 3.25 demonstrate how the adaptive rectifier adjusts its turn-on phase based on the $V_{REF}/3$ to regulate V_{REC}, with the peak voltage of $V_{INP,N}$ fixed at a 5 V peak and a carrier frequency of 2 MHz (period of 500 ns). When V_{REF} is 0.83 V, the rectifier activates 50 ns after $V_{INP,N}$ exceeds V_{REC} and turns

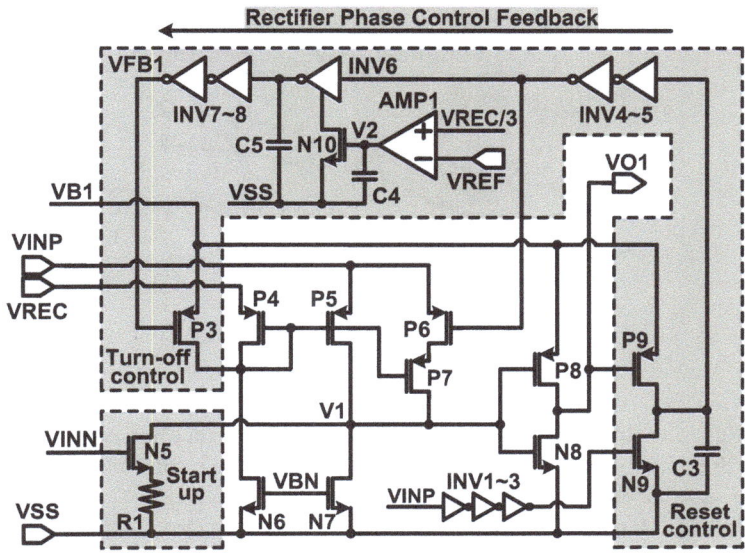

Fig. 3.24 Comparison between (**a**) conventional AC-DC rectifier and 1-stage AC-DC regulator (**b**) with turn-on time control, and (**c**) with the turn-on phase control [15]

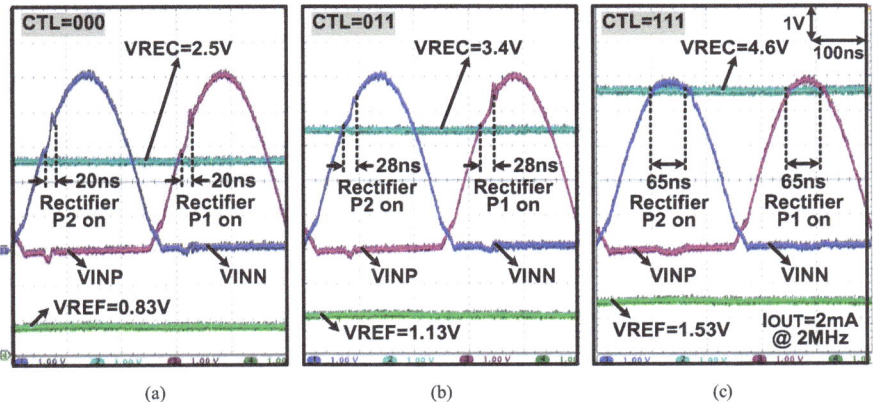

Fig. 3.25 Comparison between (**a**) conventional AC-DC rectifier and 1-stage AC-DC regulator (**b**) with turn-on time control and (**c**) with the turn-on phase control [15]

off 20 ns later, providing just enough power to raise V_{REC} to 2.5 V. For V_{REF} = 1.13 V, the rectifier turns on 66 ns, with an increased on-time of 28 ns, achieving V_{REC} = 3.4 V. When V_{REF} = 1.53 V, the rectifier functions similarly to a conventional full-wave AC-DC rectifier remaining active for 65 ns until $V_{INP,N}$ falls below V_{REC}, delivering enough power to reach a maximum V_{REC} of 4.6 V. This phase control feedback mechanism offers broad flexibility in adjusting the output to meet the specific requirements of the application.

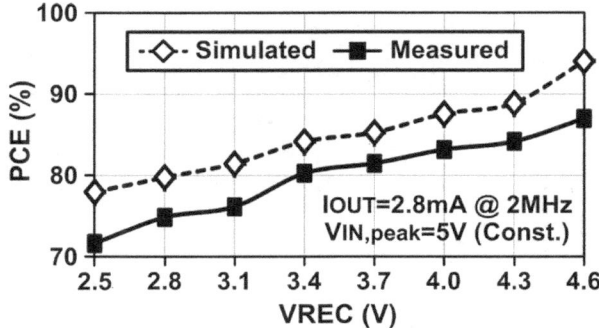

Fig. 3.26 Measured and simulated PCE with varying V_{REC} of the adaptive rectifier [15]

As illustrated in Fig. 3.26, the PCE of the adaptive rectifier is plotted against V_{REC}, with the input voltage peak ($V_{INP,N}$) fixed at 5 V and a constant load current of 2.8 mA—corresponding to the maximum stimulation current of 2.48 mA. In both simulation and experimental results, the adaptive design demonstrates strong performance, with PCE ranging from 78% to 94% in simulations and 72% to 87% in measured data. The rectifier can produce a variable output voltage from 2.5 to 4.6 V, which is set through a 3-bit digital control signal. As V_{REC} decreases, the PCE slightly drops due to two main factors: the increasing influence of dropout voltage in proportion to the output and the rising on-resistance of the switches at lower voltage levels. Despite this decline, the adaptive rectifier maintains significantly better efficiency than traditional setups where a fixed rectifier is followed by a variable linear regulator. The gap observed between measured and simulated PCE is likely caused by component mismatches—particularly between the rectifying switches and their phase detection comparators—as well as non-ideal effects such as parasitic inductance and capacitance present in the test environment.

3.3.2   Dual-Output Resonant Regulating Rectifier

The single-stage R³ design offers significant advantages in cost, size, and efficiency compared to the conventional two-stage structure, as discussed in Sect. 3.3.1. Beyond PWM R³, other techniques include topology reconfiguration modulation (TRM) R³ (Sect. 3.2.1) [16] and pulse density modulation (PDM) R³ (Fig. 3.27) [18]. TRM R³ enables mode transitions between ×1 with full-wave rectifier mode and ×N with resonant current (RCM) mode to regulate and expand the output voltage range but suffers from energy-wasted periods during RCM mode, reducing power delivered to the load (PDL) and increasing output ripple. PDM R³ regulates the output by varying the number of pulses, but unselected pulses circulating in the L_2C_2 tank dissipate power through its ESR, causing energy loss and high output ripple. Similarly, PWM R³ adjusts the gate pulse width for regulation; however,

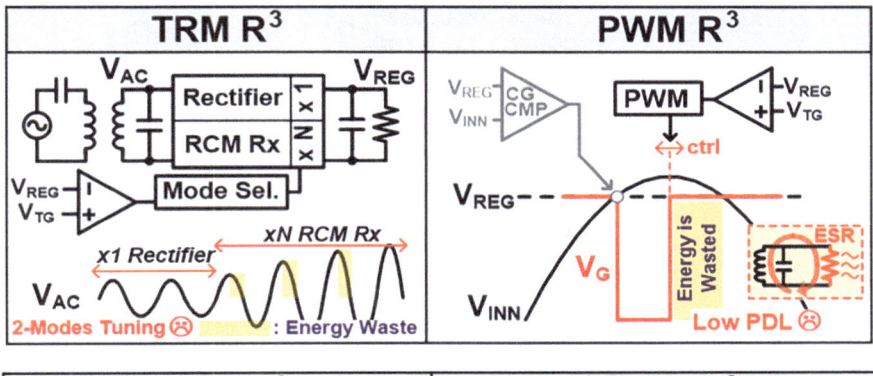

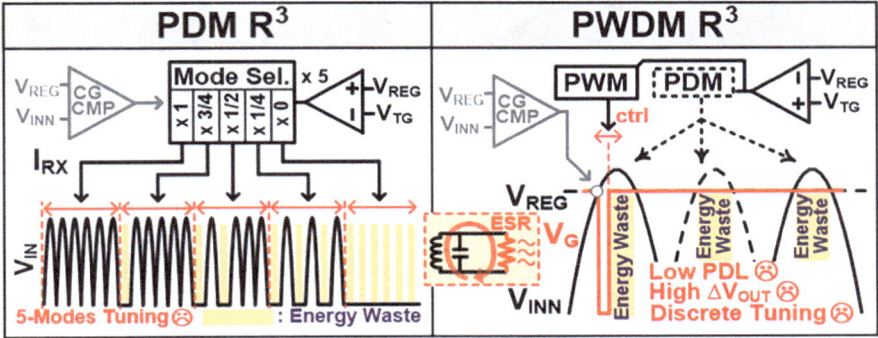

Fig. 3.27   Four types of conventional R³s [17]

residual energy dissipates through the ESR of L_2C_2 during the time between the rising edge of V_G and when V_{INN} drops to V_{REG}, resulting in heat generation and low PDL. The hybrid PWDM approach combines PWM and PDM for broader load regulation but inherits inefficiencies such as heat dissipation and limited power delivery, rendering it unsuitable for IMDs [19]. To address these limitations, the proposed energy-resuscitating R³ (ER⁴) utilizes the otherwise discarded energy to generate an additional regulated output [17]. This approach minimizes heat dissipation, enhances power delivery, and is particularly suitable for dual-output applications, such as SoCs requiring low-voltage neural recording systems and high-voltage optogenetic drivers.

The wasted energy period in the conventional PWM R³ can be reborn as the additional output V_{ER} as shown in Fig. 3.28 with its magnified waveform. The additional gate voltage (V_{G2}) is formed after the V_{G1} rises and finishes below the point as V_{INN} decreases and meets the V_{REG}. Since both V_{G1} and V_{G2} are controlled by PWM method in a half cycle during V_{INN} is larger than V_{REG}, low heat production, high PDL, and low output ripple can be achieved. A magnified waveform during 1 cycle with the detailed timing operation of the ER⁴ system features a dual-output comparator along with two PWM controllers, an offset-controlled CG comparator, a pulse-width memorizer (PWMem), and an ER-end-protector within a single V_{INN}

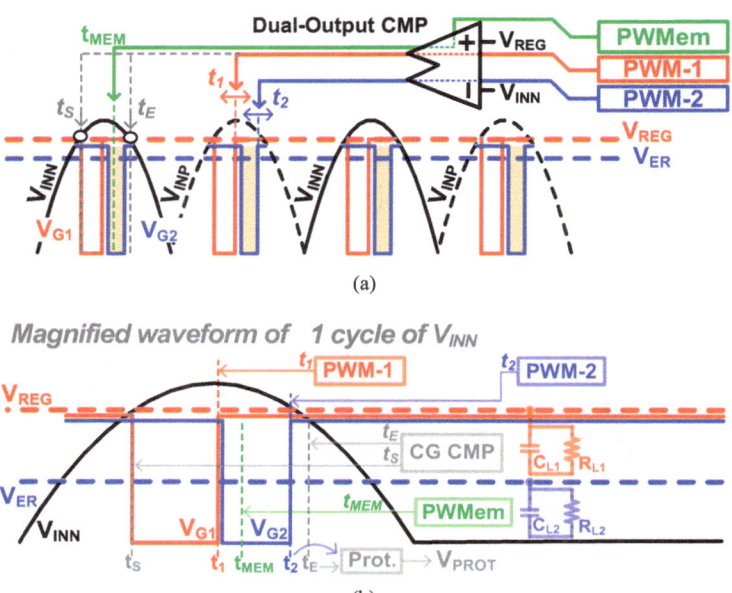

(a)

(b)

Fig. 3.28 Conceptual diagram of (**a**) proposed ER⁴ system and (**b**) its operation waveform during a single cycle [17]

cycle. The CG comparator accurately detects the moment when V_{REG} matches V_{INN}, generating V_{G1} falling edge timing (t_S) and the end-protection signal timing (t_E), focusing only on V_{REG} and V_{INN} without involving V_{ER}. V_{G1} and V_{G2} manage the energy transfer from the L_2C_2 resonant tank to two loads ($R_{L1}C_{L1}$ for V_{REG} and $R_{L2}C_{L2}$ for V_{ER}). The PWM-1 and PWM-2 controllers adjust the rising edges of V_{G1} (t_1) and V_{G2} (t_2) using analog feedback. The PWMem detects the exact t_1 where the load power reaches a preset value, storing it as t_{MEM} to guarantee sufficient time for V_{G2}. To prevent back current flow from $R_{L2}C_{L2}$ to the L_2C_2 tank when t_2 extends past t_E, the ER-end-protector generates a protective signal (V_{PROT}), ensuring t_2 does not exceed t_E.

Unlike conventional dual-output R³ designs, the proposed ER⁴ achieves regulated dual-output using only one CG comparator, the most power-intensive component of the rectifier. Instead of relying on the CG comparator to generate V_{ER}, the regulation of V_{ER} starts from t_1, improving power conversion efficiency (PCE) to 92.7% for outputs of 4.5 V at 200 Ω and 2.5 V at 1 kΩ. Additionally, by generating V_{REG} and V_{ER} within a half cycle, ER⁴ offers a wide output regulation range for V_{ER} (1.0–4.5 V) while reducing output ripple and enhancing PDL.

Two V_{GS} must remain within the region between t_S and t_E to prevent back current and should not overlap. Given the limited region and the sequential operation of V_{G1} and V_{G2}, two key considerations arise. First, V_{REG} must have a predetermined maximum value to ensure sufficient space for V_{G2}. If V_{G1} exceeds this limit, there will be no room for V_{G2}, as illustrated in Fig. 3.29a. This maximum V_{REG} is determined by

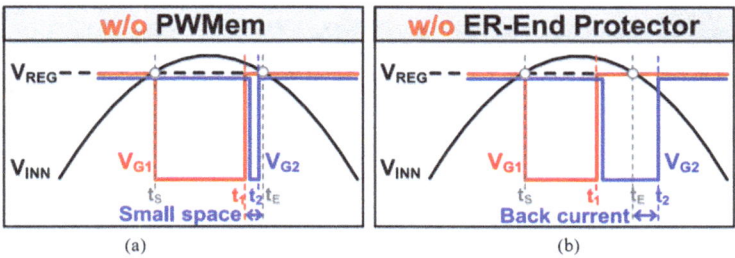

Fig. 3.29 The waveform scenarios (**a**) without PWMem and (**b**) without ER-end protector [17]

the size of the power pass PMOS transistor, and PWMem limits a maximum t_1 to prevent V_{REG} from exceeding the set value. Second, t_2 must not exceed t_E to avoid back current, as shown in Fig. 3.29b. The ER-end protector ensures that t_2 stays within this limit.

Figure 3.30a illustrates the observed output behavior of the ER^4 architecture operating in three distinct configurations: passive rectification mode (Φ_1), single-regulation mode (Φ_2), and dual-regulation mode (Φ_3), with each mode tested using two 1 μF output capacitors. In the Φ_1 state, ER^4 functions purely as a passive diode rectifier. When transitioning to Φ_2, active regulation of the primary output voltage (V_{REG}) commences and is stabilized at 4.5 V. Under this mode, the system supports only a single regulated output. Shifting further to Φ_3, ER^4 enables dual-output regulation—maintaining V_{REG} at 4.5 V while simultaneously activating a second output (V_{ER}), which is configurable to various voltage levels to 1.0 V. V_{ER} can be configured between its highest value of 4.5 V and lowest value of 1.0 V, providing a total swing of 3.5 V to suit various application needs. In Fig. 3.30b, the PCEs measured at different V_{ER} values are shown while maintaining V_{REG} at a fixed 4.5 V with a 200 Ω load. For V_{ER}, loads of 500 Ω, 750 Ω, and 1 kΩ were applied. Since V_{REG} typically provides a greater portion of the total output power, variations in V_{ER} levels between 1 V and 3.5 V have minimal impact on total efficiency, with deviations kept under 2.8%. The peak efficiency was recorded at 92.74% under conditions of 200 Ω on V_{REG} and 1 kΩ on V_{ER}, yielding an output power of 107.5 mW. Higher PCEs at increased V_{ER} loads are attributed to reduced conduction losses.

3.3.3 Multi-Output Resonant Regulating Rectifier

In the previous section, we discussed dual-output R^3 designs, but additional techniques for achieving dual-output R^3 functionality are shown in Fig. 3.31. These single-stage dual-output R^{3}s are more compact and efficient than two-stage systems combining an AC-DC rectifier with dual DC-DC converters. The first method, pulse grouping, divides cycles into two sets, each dedicated to producing either high or low regulated voltage [21]. To generate the high output voltage (V_{REG1}), PMOS

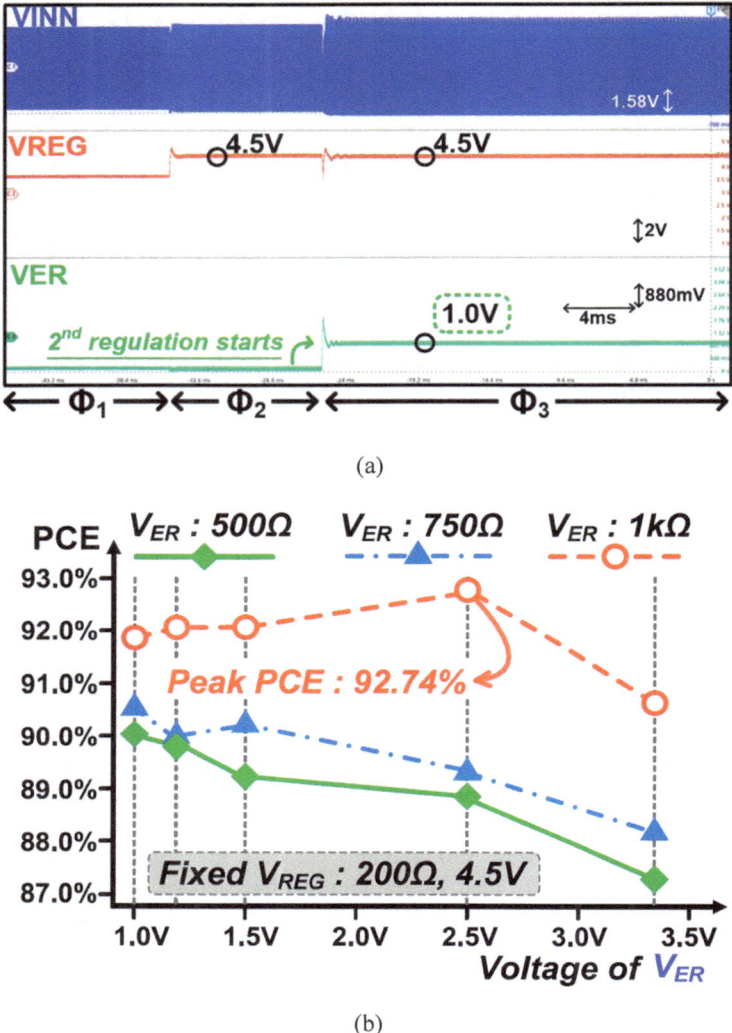

Fig. 3.30 (a) Measured waveform of V_{REG} and V_{ER}, and (b) measured PCE with different V_{ER} voltage [17]

transistors P_1 and P_2 are active for extended periods, while P_3 and P_4 remain off. Conversely, P_3 and P_4 handle the low output voltage (V_{REG2}) by being briefly active while P_1 and P_2 are off. The second approach, differential grouping, uses positive and negative cycles to create two separate outputs. P_5 transfers energy from V_{INN} to V_{REG1}, while N_0 supplies power from V_{INP} to V_{REG2} [22]. The third technique, called the middle outlet method, employs a voltage doubler to produce V_{REG1} and derives V_{REG2} from the midpoint of the doubler [23]. This can also be viewed as a variation

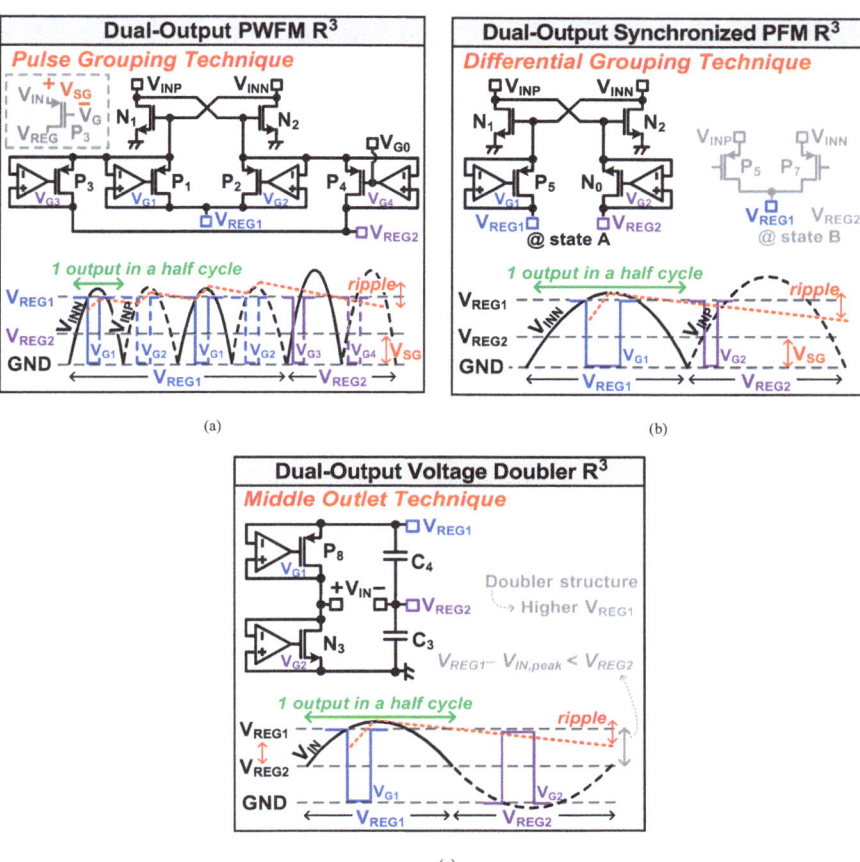

Fig. 3.31 Three types of conventional dual-output resonant regulating rectifier (R^3) techniques, which are (**a**) dual-output PWFM R^3, (**b**) dual-output synchronized PFM R^3, and (**c**) dual-output voltage doubler R^3 [20]

of differential grouping. Since all three techniques enable one output per half cycle, they result in high output ripple and reduced power delivered to the load (PDL). Additionally, conventional designs often face challenges with low efficiency and limited output regulation range for the low-voltage output (V_{REG2}).

Figure 3.32 illustrates the single-input multi-output (SIMO) R^3 system, capable of producing three outputs within a half cycle using distributed multi-phase control [20]. The half cycle is divided into three regions—peak left (PL), peak center (PC), and peak right (PR)—to ensure the three V_Gs do not overlap. Protection techniques are applied within each region to maintain separation. The highest voltage ($V_{REG,PC}$) is generated in the PC region and powers the blocks in the PL and PR regions, which then produce the corresponding voltages ($V_{REG,PL}$ and $V_{REG,PR}$). This distributed multi-phase control in a half cycle reduces output ripple and enhances PDL. To

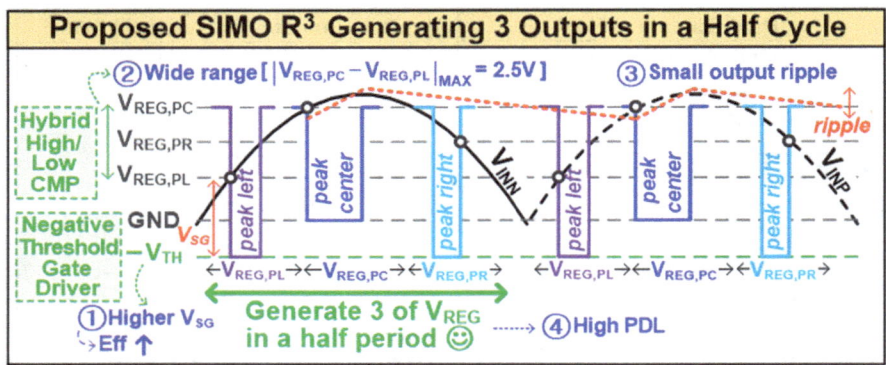

Fig. 3.32 Conceptual diagram of the SIMO R³ generating three outputs in a half cycle [20]

further improve efficiency and expand the regulation range of secondary outputs, a hybrid high/low comparator (HCMP) and a negative threshold gate driver (NTGD) are introduced, as shown in Fig. 3.32.

Figure 3.33a shows HCMP, a hybrid comparator that integrates PMOS-input and NMOS-input CG comparators to widely detect the point where V_{INN} equals $V_{REG,PR}$. HCMP operates as a PMOS-input CG comparator for identifying when $V_{REG,PR}$ matches the upper range of V_{INN} and acts as an NMOS-input CG comparator for detecting when $V_{REG,PR}$ aligns with the lower range of V_{INN}. This technique can lead to wide regulation range in the secondary output. Figure 3.33b presents NTGD, designed to lower $V_{REG,PR}$ to $-V_{TH}$, thereby increasing the source-to-gate voltage (V_{SG}) of power pass PMOS transistor in PR region (P_{PR}). This increase in V_{SG} offers two main benefits: a lower R_{ON} for the P_{PR} transistor, enhancing power conversion efficiency (PCE) during $V_{REG,PR}$ regulation, and an expanded regulation range for $V_{REG,PR}$ by allowing P_{PR} to operate in the triode region even at lower V_{INN}. The NTGD is built using three key components: a diode-connected PMOS (D_1), a threshold voltage storage capacitor (C_3), and a current-limiting resistor (R_1). NTGD functions in two phases as $V_{G,PR}$ transitions. When $V_{G,PR}$ rises, it connects to $V_{REG,PC}$, turning P_{PR} off. Current flows from $V_{REG,PC}$ to ground through P_{13}, D_1, P_{16}, and R_1, charging C_3 to the threshold voltage (V_{TH}) of D_1. As $V_{G,PR}$ falls, the circuit shifts to a path involving N_7, C_3, and N_9, lowering $V_{G,PR}$ to $-V_{TH}$. This switching mechanism allows $V_{G,PR}$ to alternate between $V_{REG,PC}$ and $-V_{TH}$, distinguishing NTGD from conventional gate drivers in dual-output R³ systems.

In the PL and PC regions, forward PWM control is used for its straightforward design, but in the PR region, reverse PWM control is crucial for improving efficiency. Figure 3.34 compares these approaches within the PR region. Under forward control, PWM controller sets the duration of $V_{G,PR}$ starting from rising timing of $V_{G,PC}$ ($t_{E,PC}$), but this leads to significant conduction losses (P_{COND}) in P_{PR} due to the large voltage difference between V_{INN} and $V_{REG,PR}$, ultimately lowering PCE. In contrast, reverse control initiates $V_{G,PR}$ from the rising timing of $V_{G,PR}$ ($t_{E,PR}$), with

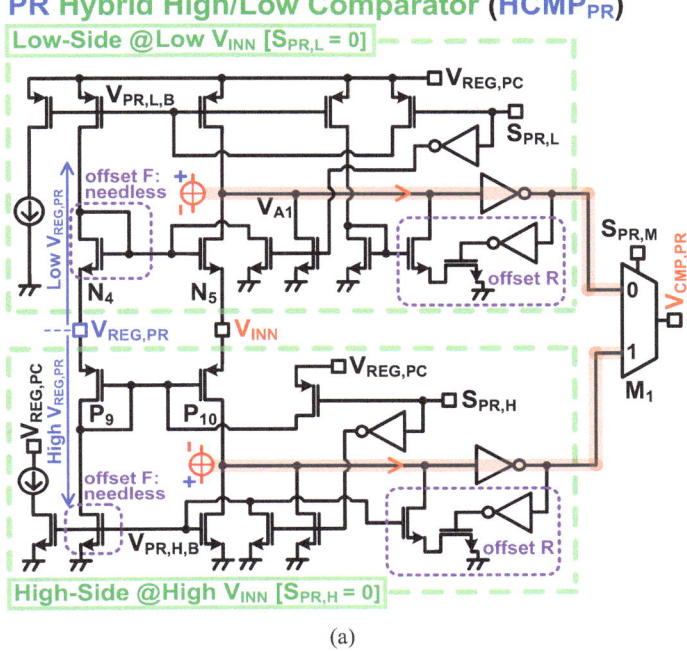

(a)

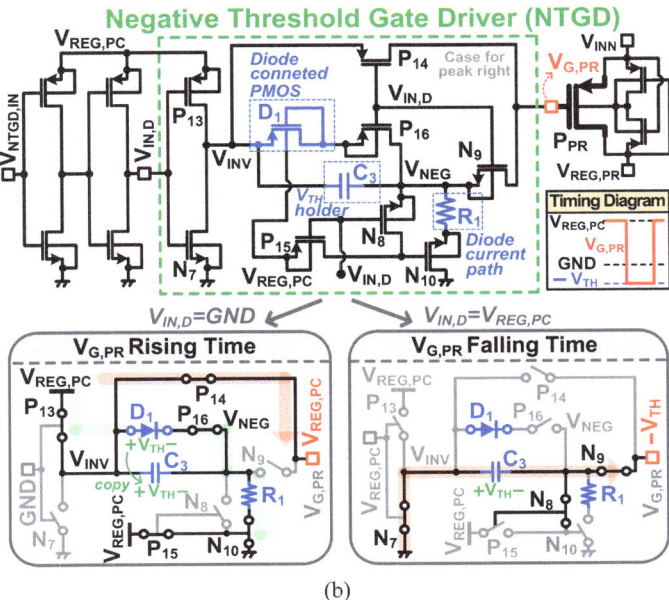

(b)

Fig. 3.33 Circuit techniques to improve the (**a**) $V_{REG,PR}$ regulation range with hybrid high/low comparator (HCMP) and (**b**) efficiency of $V_{REG,PR}$ with negative threshold gate driver (NTGD) [20]

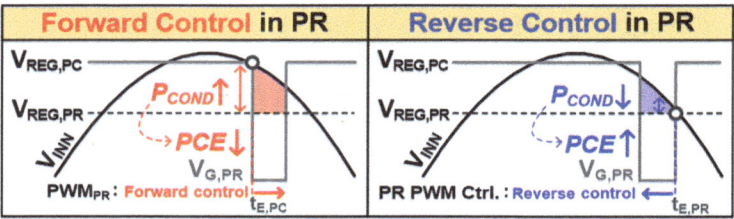

Fig. 3.34 Conceptual diagram comparing forward and reverse control in PR [20]

PWM adjusting its duration. This approach minimizes the voltage difference between V_{INN} and $V_{REG,PR}$, significantly enhancing PCE.

The distributed multi-phase control in the SIMO R^3 divides operations into three regions—PL, PC, and PR—to generate three independent V_{GS}. In the PL and PR regions, V_{INN} does not reach its maximum but instead rises or falls, necessitating the use of HCMP and NTGD for precise control. Additionally, three pairs of CG comparators are required since each region independently manages its respective V_G. In contrast, the ER^4 system, discussed in the previous Section, generates two V_{GS} within the PC region, reducing the need to just one pair of CG comparators. However, ER^4 faces a trade-off: as the difference between V_{INN} and the secondary output voltage increases, the efficiency of the secondary output decreases. Despite this, the ER^4 offers a wider regulation range for the secondary output compared to SIMO R^3, as it is generated in the PC region. Thus, applications must weigh the trade-off between the secondary output regulation range and its PCE when choosing between these architectures.

Figure 3.35a demonstrates the functionality of the SIMO R^3 system in producing three distinct output voltages within a single half cycle, enabled by a distributed multi-phase control scheme. The power conversion process begins with the activation of the P_{PC} transistor, which stabilizes $V_{REG,PC}$ at 4.5 V. This regulated output is then routed to the left and right regulation paths—$TCMP_{PL}$ and $TCMP_{PR}$—which drive P_{PL} and P_{PR} to establish $V_{REG,PL}$ and $V_{REG,PR}$ at 3.3 V and 1.8 V, respectively. SIMO R^3 supports significant output flexibility, as both $V_{REG,PL}$ and $V_{REG,PR}$ can be tuned over a wide voltage range, from a maximum of 3.5 V down to a minimum of 1.0 V.

Figure 3.35b presents efficiency measurements taken across different $V_{REG,PL}$ levels, with the load fixed at 1 kΩ, under two output configurations referred to as case (1) and case (2). A positive correlation is observed between rising $V_{REG,PL}$ values and PCE, primarily due to a reduction in switch resistance (R_{ON}) as the gate-source voltage (V_{SG}) of P_{PL} increases. SIMO R^3 supports low-voltage operation down to 1.0 V for $V_{REG,PL}$, made possible through the combined roles of the $HCMP_{PL}$ circuit—which detects when $V_{REG,PL}$ matches a low input level—and the NTGD driver, which shifts $V_{G,PR}$ to below the minus threshold voltage ($-V_{TH}$). Since the output power in case (1) is approximately 1.52 times greater than in case (2), its overall efficiency is also higher. The highest recorded efficiency of 90.82% occurs when the system delivers a total output of 84.6 mW, with $V_{REG,PC}$ at 4.5 V across a 300 Ω load, $V_{REG,PR}$ at 3.3 V across 1 kΩ, and $V_{REG,PL}$ at 2.5 V across 1 kΩ.

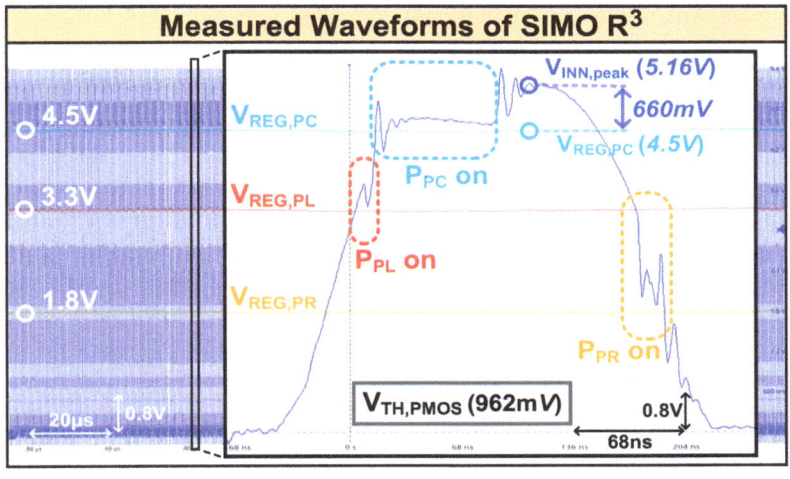

(a)

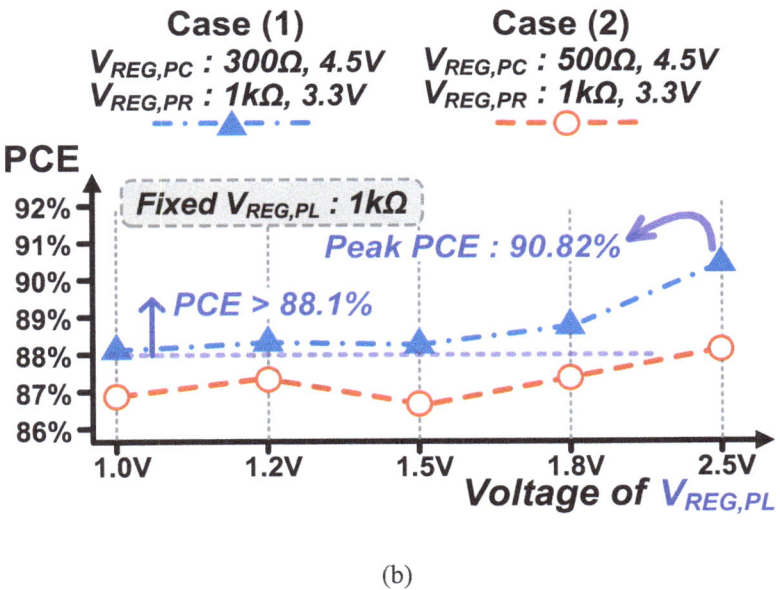

(b)

Fig. 3.35 (**a**) Measured waveform of SIMO R³ generating 3 outputs in a half cycle and (**b**) measured PCEs with varying $V_{REG,PL}$ voltage conditions [20]

3.4   Resonant-Mode Energy Receiver

3.4.1   Time-Interleaved Resonant-Mode Energy Receiver

The inductive link-based WPT systems discussed in Sects. 3.1, 3.2, and 3.3 operate effectively in the near field, typically within a few centimeters. However, for IMD applications located deeper in the body, with link distances of 7–8 cm, the receiver can only generate a small voltage. To address this limitation, resonant current mode (RCM) receivers have been developed. These receivers accumulate small amounts of input energy over several resonance cycles and deliver it to the load as a boosted current. In this section, we will explore the RCM Rx, resonant voltage mode (RVM) Rx, and non-residual energy Rx systems.

Both RCM Rx and RVM Rx systems share two operational phases: an accumulation phase and a transfer phase. The key distinction lies in when the transfer phase begins. In RCM Rx, the transfer phase starts when the energy stored in the inductor reaches its peak, while in RVM Rx, it begins when the energy stored in the capacitor is at its maximum, as illustrated in Fig. 3.36 [24]. In RCM Rx, the resonant current (I_{AC}) peaks when the resonant voltage (V_{RX}) drops to zero. Detecting this zero-voltage point requires a zero-voltage detector, typically implemented with a high-speed comparator. However, achieving precise detection or operating at higher resonant frequencies demands a comparator with a higher bias current. Conversely, in RVM Rx, the transfer phase starts when V_{RX} reaches its maximum and I_{AC} falls to zero [25]. This zero-current point is detected using a passive diode-based zero-current detector, which allows for faster operation, especially at higher resonant frequencies.

In RVM Rx, a passive diode pre-biased to $-V_{TH}$ is used to detect zero current, identifying the point where the capacitor voltage (V_{AC}) reaches its maximum (V_{PK}),

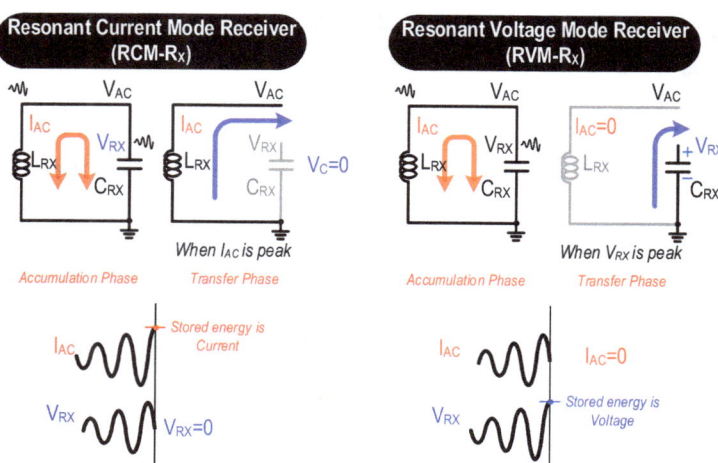

Fig. 3.36 Conceptual diagrams of RCM Rx and RVM Rx

as shown in Fig. 3.37. During multiple resonance cycles, the resonant switch (S) remains closed, functioning as an on-resistance (R_{ON}) to form the LC resonant tank. After these cycles, the switch transitions to a passive diode with the same pre-biased $-V_{TH}$. When V_{AC} peaks at V_{PK} and the resonant current (I_{AC}) drops to zero, the diode passively turns off. At this point, V_{PK} is stored across the resonant capacitor (C_{RX}) and the parasitic capacitance (C_P). Due to the significantly smaller size of C_P compared to C_{RX}, V_{AC} rapidly decreases from V_{PK} once the diode turns off. This sharp decline in V_{AC} indicates the peak voltage point and marks the transition from the resonant phase to the charging phase.

The proposed RVM Rx employs a time-interleaved mode utilizing two capacitors, as illustrated in Fig. 3.38. During phase Φ_1, C_{RX1} forms an LC resonant tank with L_{RX}, while C_{RX2} transfers energy to the battery through the quasi-resonant boost converter (QRBC). In phase Φ_2, the roles reverse—C_{RX2} pairs with L_{RX} to create the

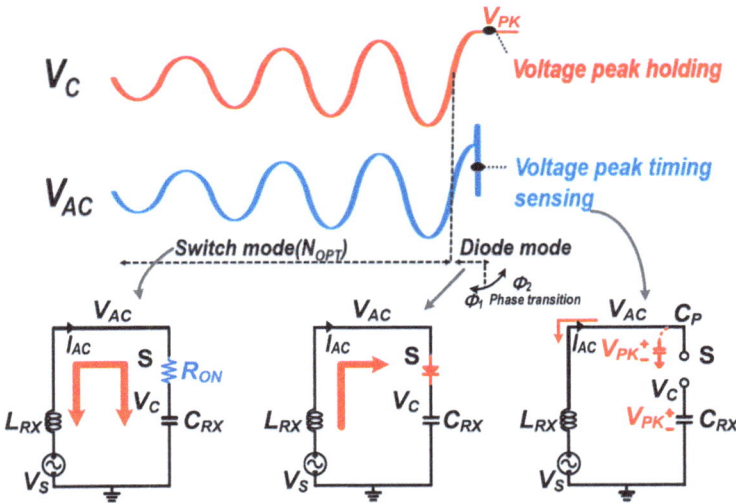

Fig. 3.37 Conceptual diagram of RVM Rx detects the resonant inductor current (I_{AC}) becomes zero [25]

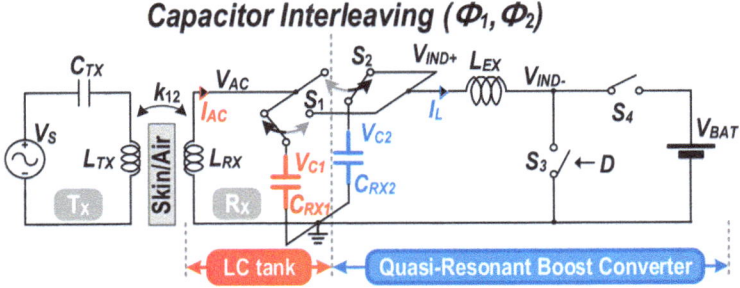

Fig. 3.38 Proposed time-interleaved scheme in RVM Rx [25]

LC resonant tank, while C_{RX1} facilitates energy transfer to the battery via the QRBC. This time-interleaved operation ensures a continuous energy flow from the Tx, regardless of variations in the load, thereby maintaining a consistent power transfer efficiency (PTE).

Figure 3.39 illustrates the fundamental operation of the time-interleaved RVM Rx system, which operates in two alternating phases (Φ_1 and Φ_2). These phases are determined by the optimal number of resonant cycles (N_{OPT}) needed to amplify V_{AC} based on the input power of receiver (P_{RX}). The system transitions between phases when the voltage across the resonant capacitor (V_C—V_{C1} in Φ_1 or V_{C2} in Φ_2) reaches its maximum after completing N_{OPT} cycles. At the transition point, all energy in the LC tank is concentrated in the resonant capacitor (C_{RX1} or C_{RX2}) and is then transferred to the battery via the QRBC. This time-interleaved design completely isolates the LC tank from the output, enabling the RVM Rx to continuously receive P_{RX}

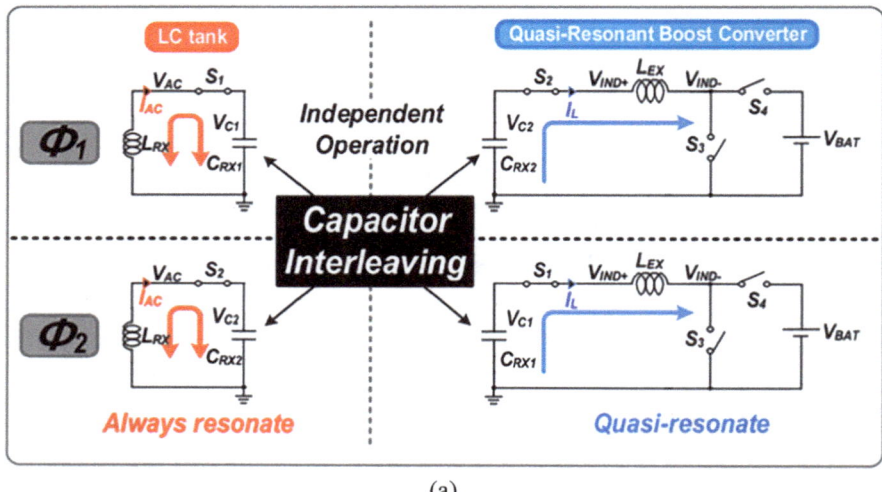

(a)

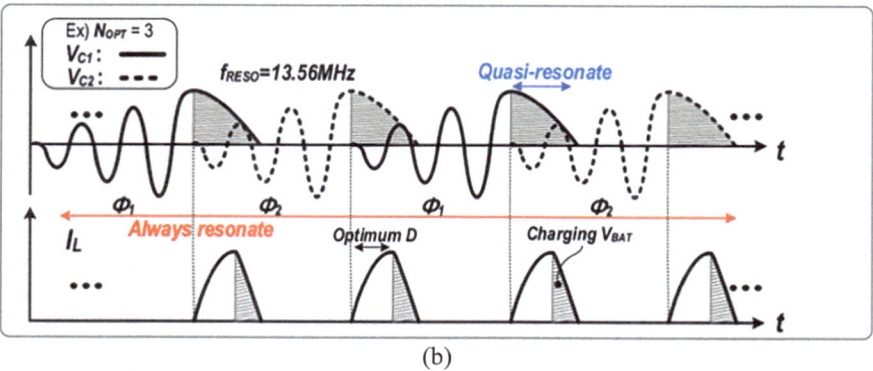

(b)

Fig. 3.39 (a) Time-interleaved operation phases and (b) its timing diagram [25]

from the Tx regardless of load fluctuations, such as variations in V_{BAT}. Furthermore, the system maintains a stable LC tank configuration when receiving wireless power, ensuring uninterrupted resonance, and optimizing P_{RX} reception.

Figure 3.40a presents experimental waveforms of the RVM-based receiver operating under steady-state conditions. At a received power (P_{RX}) of 9.5 mW and N_{OPT}

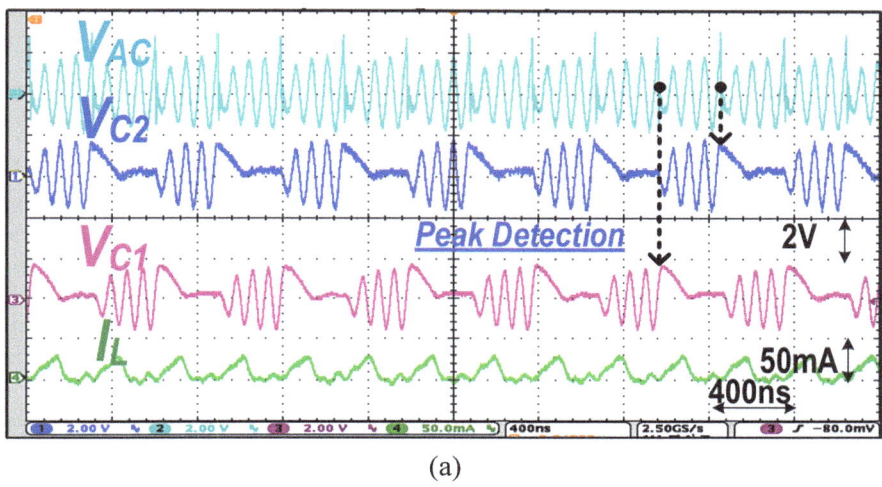

(a)

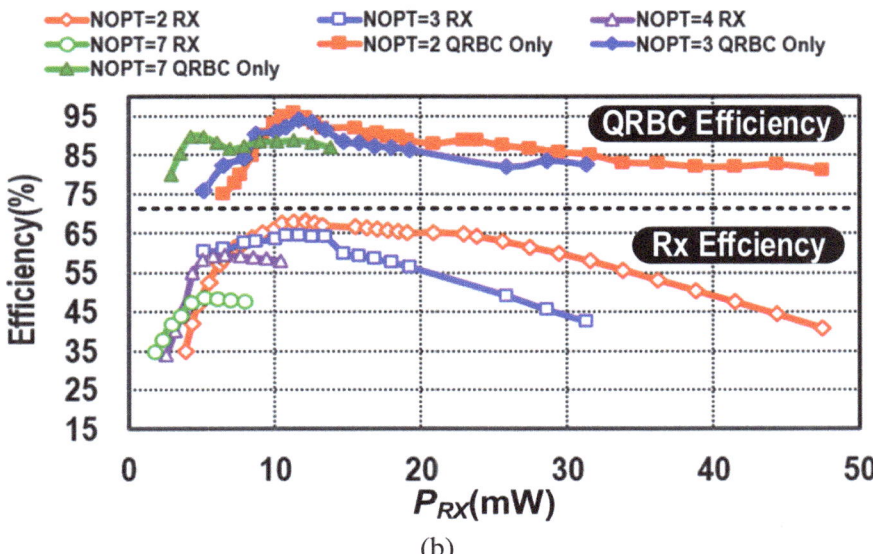

(b)

Fig. 3.40 (a) Measured steady-state waveforms with N_{OPT} of 4 and (b) measured PCE of overall Rx and QRBC [25]

of 4, the charging behavior of capacitors V_{C1} and V_{C2} demonstrates alternating energy delivery to the battery. During this process, the QRBC efficiently draws the accumulated energy from the resonant capacitor and transfers it to the battery. Figure 3.40b presents the PCEs of both Rx and QRBC across different input power levels. At lower input powers, the Rx achieves higher PCE due to reduced switching losses, especially as N_{OPT} increases. When N_{OPT} is set to 2, peak efficiencies are measured at 67.8% for the Rx and 95% for the QRBC.

3.4.2   Non-Residual Resonant-Mode Energy Receiver

In conventional RCM Rx or RVM Rx systems, the energy stored in the L_2C_2 tank is not entirely transferred to the battery. Residual energy remains circulating within the L_2C_2 tank, where it dissipates as heat due to the equivalent series resistance (ESR) of the components. This inefficiency results in reduced power transfer efficiency (PTE) and lower power delivered to the load (PDL). Additionally, the parallel-LC structure in RCM Rx systems cannot handle voltages exceeding the transistor breakdown threshold, posing a risk of IC damage. This limitation restricts the range of input voltages the system can accommodate, further constraining PDL.

To address these problems, arise from the non-residual energy in the L_2C_2 tank and the limited input voltage range due to parallel-LC resonant tank, the series-LC non-residual RCM Rx is proposed in Fig. 3.41 [26]. To perform the non-residual energy after charging period finishes, the energy in the L_2 and C_2 has to be zero at the end of the charging period. Thus, the current in the L_2 (I_{L2}) and the voltage in the C_2 (V_{C2}) have to be monitored with the proposed controller, named as dual automated maximum efficiency control (AMEC) and power delivery should be done until these values become zero. Also, to enlarge the input voltage range, the series

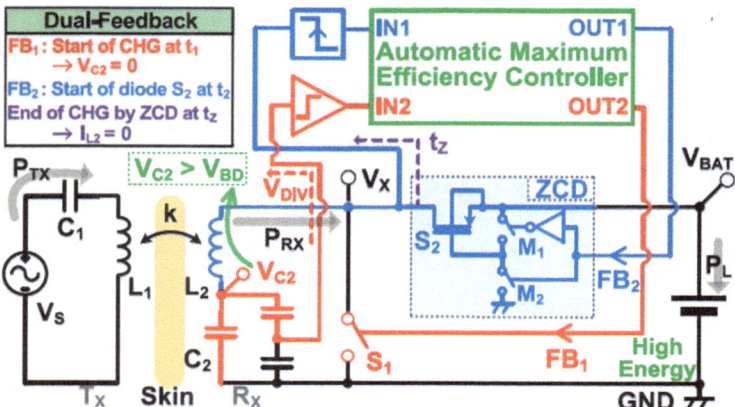

Fig. 3.41 Conceptual diagrams of the proposed series-LC RCM Rx [26]

LC tank in the receiver should be adopted, thus the V_{C2} can be much larger than the transistor breakdown voltage.

The series-LC RCM Rx operates in two distinct modes. (1) Resonant mode: S_1 connects V_X to GND, boosting V_{C2} received from the T_X while S_2 remains open. (2) Charging mode: S_1 opens while S_2 connects V_X to V_{BAT}, transferring energy from the L_2C_2 tank to the battery. As the transfer finishes, S_2 transitions into diode mode, ensuring all residual energy in the L_2C_2 tank is delivered to the battery until the residual energy reaches zero. The dual AMEC employs two feedback signals, FB_1 and FB_2, to maximize energy transfer. FB_1 monitors V_{C2} to determine the optimal end-of-resonance timing (t_1) which is also the start of charging. This ensures V_{C2} reaches zero by the end of the charging phase. Simultaneously, the ZCD passively switches S_2 to diode mode when the current through S_2 (I_{S2}) drops to zero at t_Z, which is also end-of-charging timing and the start of the next resonance cycle without idle time. FB_2 detects t_Z and calculates the optimal timing (t_2) for diode conversion of S_2, ensuring safe and efficient operation. In summary, the proposed series-LC RCM Rx effectively manages energy transfer, charging the battery until V_{C2} and I_{S2} reach zero at the end of each charging cycle, ensuring high efficiency and reliability.

Furthermore, the voltage stress on switches (S_1 and S_2) is limited to V_{BAT}, staying below the V_{BD}. In resonant mode, S_1 connects V_X to GND, while in charging mode, S_2 connects V_X to V_{BAT}. Thus, even if V_{C2} exceeds V_{BD} to enlarge the input voltage range, the system ensures safe operation by keeping the maximum voltage across the on-chip switches below V_{BD}.

The series-LC non-residual RCM Rx operates in three phases as shown in Fig. 3.42. Phase Φ_1 represents the resonant mode, whereas phases Φ_2 and Φ_3 are charging modes. During phase Φ_1, S_1 closes and S_2 is opened by closing M_1 and opening M_2 in ZCD, then V_X connects to GND. The L_2C_2 tank undergoes multiple resonant cycles, increasing V_{C2}. At the optimal time t_1 triggered by FB_1, S_1 is opened and the phase Φ_2 which is the initial energy transfer step begins. Also, in the ZCD, M_1 opens and M_2 closes, converting S_2 into a closed switch. As a result, V_X connects to V_{BAT}, transferring the stored energies in the L_2 and C_2 to the battery. The timing t_1 is optimized to ensure V_{C2} reaches zero at the end of charging at t_Z. At the optimal time t_2, determined by FB_2, the phase Φ_3 which is the final energy transfer step starts. In the ZCD, M_1 closes and M_2 opens, reconfiguring S_2 into a parasitic body diode in parallel, transferring the remaining energy from L_2 and C_2 to the battery. This process continues until I_{S2} drops to zero, causing V_X to fall from V_{BAT} to GND. At this point, the parasitic diode operation ends and S_2 becomes an open switch at the end of charging at t_Z. The ZCD passively tracks t_Z, ensuring I_{S2} reaches zero. Once t_Z is reached, the system immediately transitions back to phase Φ_1, initiating resonance in V_{C2} without any idle period. This seamless transition eliminates idle time between charging and resonant modes, maximizing the charging duration relative to the total cycle time and increasing power delivery efficiency (PDL). In conclusion, the dual AMEC ensures complete energy transfer from the L_2C_2 tank to the battery, leaving no residual energy in L_2 or C_2 at the end of charging, thus increasing both PDL.

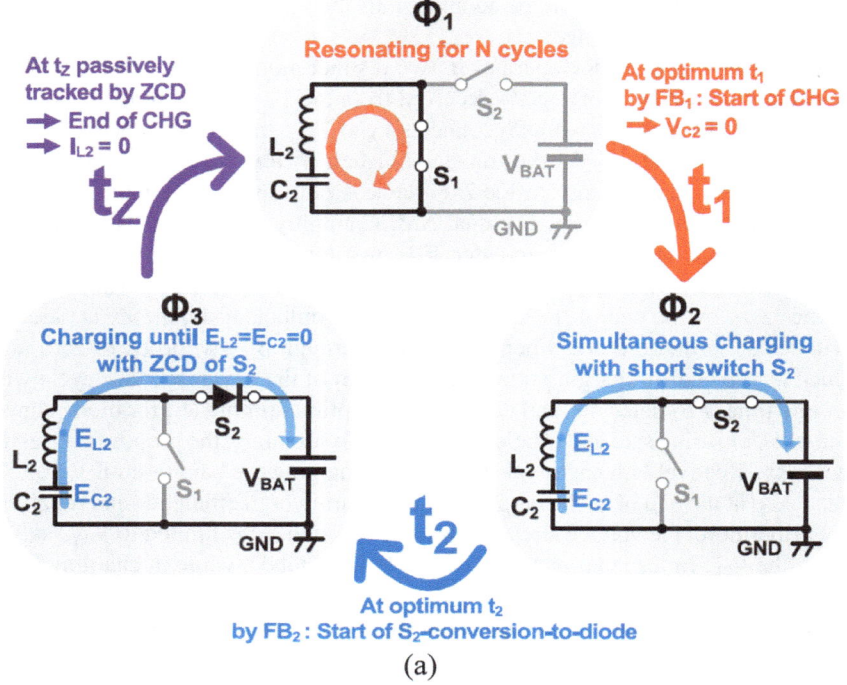

(a)

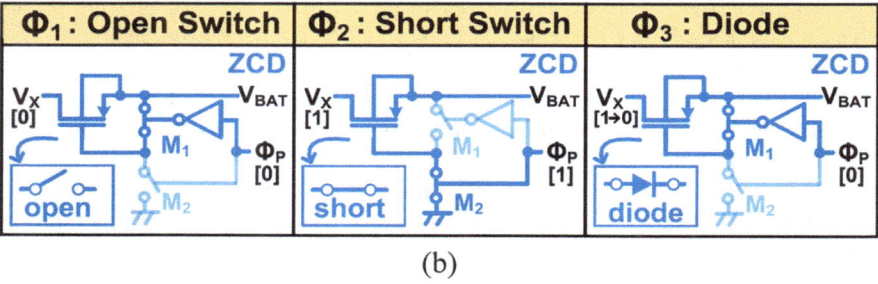

(b)

Fig. 3.42 (**a**) Conceptual diagrams of operation phases and (**b**) ZCD configurations [26]

Figure 3.43a displays the measured waveforms for $N = 3$, emphasizing the precise end-of-charging point where both V_{C2} and I_{L2} reach zero, achieved through the proposed dual AMEC and passive ZCD mechanisms. The two digital feedbacks of dual AMEC loops dynamically adjust to identify the optimal charging start time (t_1) and the ideal moment for S_2 to switch to diode mode (t_2). Meanwhile, the passive ZCD system accurately detects the end-of-charging timing (t_Z). As a result, the L_2C_2 tank is left with no residual energy at the end of charging, ensuring high PDL. This system differs from both RCM Rx and RVM Rx described in the previous section, as the transfer phase begins somewhere between the maximum

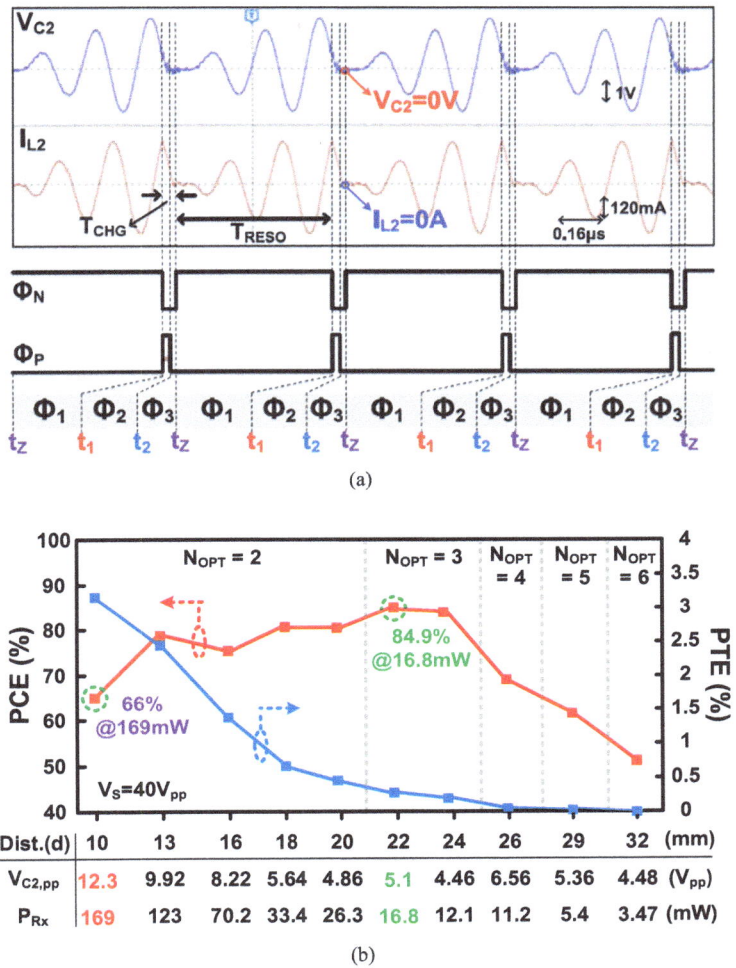

Fig. 3.43 (**a**) Measured waveforms of V_{C2} and I_{L2} (or I_{S2}) showing non-residual energy at the end of the charging mode and (**b**) measured PCEs with varying distance [26]

energy accumulation in the coil or capacitor. If a name were to be assigned, it could be called the resonant half-voltage half-current mode Rx. Figure 3.43b presents experimental data illustrating PCE vary with respect to the coil separation distance (d), the peak-to-peak voltage of the resonant node ($V_{C2,pp}$), and the received power at the Rx (P_{RX}), under a constant transmitter voltage of 40 V_{pp} applied to the primary coil L_1. When the coil spacing was set to 10 mm and the battery voltage V_{BAT} was 1.8 V, the receiver achieved a maximum $V_{C2,pp}$ of 12.32 V, with P_{RX} reaching 169 mW.

3.5 Conclusions

This chapter explored the evolution of wireless energy management ICs, from basic passive rectifiers to advanced adaptive, multi-output, and resonant-mode receivers. Active rectification techniques with offset-control, voltage multipliers, and optimal-tracking algorithm have significantly improved PCE and voltage regulation. Furthermore, dual- and multi-output regulating rectifiers and resonant-mode archi-tectures enable efficient power delivery across a wide range of conditions. Innovations such as energy-resuscitating scheme and non-residual RCM receivers demonstrate how circuit-level intelligence can maximize power transfer efficiency while reducing ripple, area, and thermal loss.

References

1. P. Li, R. Bashirullah, A wireless power interface for rechargeable battery operated medical implants. IEEE Trans. Circuits Syst. II **54**(10), 912–916 (2007)
2. G. Bawa, M. Ghovanloo, Active high power conversion efficiency rectifier with built-in dual-mode back telemetry in standard CMOS technology. IEEE Trans. Biomed. Circuits Syst. **2**(3), 184–192 (2008)
3. T.T. Le, J. Han, A. von Jouanne, K. Mayaram, T.S. Fiez, Piezoelectric micro-power generation interface circuits. *IEEE J. Solid State Circuits* **41**(6), 1411–1420 (2006)
4. S. Guo, H. Lee, An efficiency-enhanced CMOS rectifier with unbalanced- biased compara-tors for transcutaneous-powered high-current implants. IEEE J. Solid State Circuits **44**, 1796–1804 (2009)
5. H.-M. Lee, M. Ghovanloo, An integrated power-efficient active rectifier with offset-controlled high speed comparators for inductively powered applications. IEEE Trans. Circuits Syst. I, Reg. Papers **58**(8), 1749–1760 (2011)
6. M. Ghovanloo, K. Najafi, Fully integrated wideband high-current rectifiers for inductively powered devices. IEEE J. Solid-State Circuit **39**(11), 1976–1984 (2004)
7. H.-M. Lee, M. Ghovanloo, A high frequency active voltage doubler in standard CMOS using offset-controlled comparators for inductive power transmission. IEEE Trans. Biomed. Circuits Syst. **7**(3), 213–224 (2013)
8. H.-M. Lee, C.S. Juvekar, J. Kwong, A.P. Chandrakasan, A nonvolatile flip-flop-enabled cryp-tographic wireless authentication tag with per-query key update and power-glitch attack coun-termeasures. IEEE J. Solid State Circuits **52**(1), 272–283 (2017)
9. H.-G. Rhew, J. Jeong, J.A. Fredenburg, S. Dodani, P.G. Patil, M.P. Flynn, A fully self-contained logarithmic closed-loop deep brain stimulation SoC with wireless telemetry and wireless power management. IEEE J. Solid State Circuits **49**(10), 2213–2227 (2014)
10. H.-M. Lee, M. Ghovanloo, An adaptive reconfigurable active voltage doubler/rectifier for extended-range inductive power rransmission. *IEEE Trans. Circuits Syst. II* **59**(8), 481–485 (2012)
11. L. Cheng, W. Ki, Y. Lu, T. Yim, Adaptive on/off delay compensated active rectifiers for wire-less power transfer systems. IEEE J. Solid-State Circuit **51**(3), 712–723 (2016)
12. C. Huang, T. Kawajiri, H. Ishikuro, A near-optimum 13.56-MHz CMOS active rectifier with circuit-delay real-time calibrations for high-current biomedical implants. IEEE J. Solid-State Circuit **51**(8), 1797–1809 (2016)
13. J. Ahn, H.-S. Lee, K. Eom, W. Jung, H.-M. Lee, A 13.56-MHz 93.5%-efficiency optimal on/off timing tracking active rectifier with digital feedback-based adaptive delay control. *IEEE Trans. Biomed. Circuits Syst.* **19**(3), 562–576 (2025)

14. H.-S. Lee, J. Ahn, M. Kang, H.-M. Lee, A load-insensitive hybrid lsk back telemetry system with slope-based demodulation for inductively powered biomedical devices. IEEE Trans. Biomed. Circuits Syst. **16**(4), 651–663 (2022)

15. H.-M. Lee, H. Park, M. Ghovanloo, A power-efficient wireless system with adaptive supply control for deep brain stimulation. IEEE J. Solid-State Circuit **48**(9), 2203–2216 (2013)

16. H. Sadeghi Gougheri, M. Kiani, Self-regulated reconfigurable voltage/current-mode inductive power management. IEEE J. Solid-State Circuit **52**(11), 3056–3070 (2017)

17. H.-S. Lee, H.-M. Lee, A 92.7%-efficiency 6.78-mhz energy-resuscitating resonant regulating rectifier with dual outputs for wirelessly powered devices. IEEE J. Solid-State Circuit **59**(10), 3204–3217 (2024)

18. J. Lin, C. Zhan, Y. Lu, A 6.78-MHz single-stage wireless power receiver with ultrafast transient response using hysteretic control and multilevel current-wave modulation. *IEEE Trans. Power Electron.* **36**(9), 9918–9926 (2021)

19. C. Kim, S. Ha, J. Park, A. Akinin, P.P. Mercier, G. Cauwenberghs, A 144-MHz fully integrated resonant regulating rectifier with hybrid pulse modulation for mm-sized implants. *IEEE J. Solid State Circuits* **52**(11), 3043–3055 (2017)

20. H.-S. Lee, K. Eom, H.-M. Lee, A single-input multi-output resonant regulating rectifier generating three outputs in a half cycle for wirelessly powered biomedical devices. *IEEE Trans. Circuits Syst. I, Reg. Papers.* **72**(6), 2956–2969 (2025)

21. R. Erfani, F. Marefat, P. Mohseni, A dual-output single-stage regulating rectifier with PWM and dual-mode PFM control for wireless powering of biomedical implants. *IEEE Trans. Biomed. Circuits Syst.* **14**(6), 1195–1206 (2020)

22. Z. Luo, J. Liu, H. Lee, A high-efficiency 40.68-MHz single-stage dual-output regulating rectifier with ZVS and synchronous PFM control for wireless powering. *IEEE J. Solid State Circuits* **59**(8), 2418–2429 (2024)

23. T. Lu, K.A.A. Makinwa, S. Du, A single-stage dual-output regulating voltage doubler for wireless power transfer. *IEEE J. Solid State Circuits* **59**(9), 2922–2933 (2024)

24. M. Choi, T. Jang, J. Jeong, S. Jeong, D. Blaauw, D. Sylvester, A current-mode wireless power receiver with −32 dBm sensitivity for implantable systems. *IEEE J. Solid State Circuits* **52**(12), 2880–2892 (2016)

25. S.-U. Shin, M. Choi, S. Jung, H.-M. Lee, G.-H. Cho, A time-interleaved resonant voltage mode wireless power receiver with delay-based tracking loops for implantable medical devices. *IEEE J. Solid State Circuits* **55**(5), 1374–1385 (2020)

26. H.-S. Lee, J. Ahn, K. Eom, W. Jung, S.-J. Lee, Y.-W. Jung, S.-U. Shin, H.-M. Lee, A power-efficient resonant current mode receiver with wide input range over breakdown voltages using automated maximum efficiency control. *IEEE Trans. Power Electron.* **37**(7), 8738–8750 (2022)

Chapter 4
Wireless Data Telemetry

4.1 Introduction

Over the past few decades, wireless applications have been developed in various industrial fields such as portable devices, Internet-of-Things (IoT), and biomedical systems [1]. As electronic devices continue to shrink in size, these applications often incorporate small batteries, weakly coupled coils, and miniature coils and antennas [2]. Although wireless power transfer (WPT) technology has been adopted as a method to either replace or recharge batteries, the limited size of the receiver (Rx) coil in wireless transmission devices and the short distance of a few centimeters between the transmitter (Tx) and Rx coils result in relatively low-power transfer efficiency (PTE). In such scenarios, issues such as temperature rise and electromagnetic interference can occur, making it crucial to minimize power consumption [3].

Figure 4.1 illustrates the average power consumption of low-noise amplifiers, signal conditioning, digitization, processing, and RF data communication blocks in biomedical wireless applications [4]. These wireless systems typically employ far-field data communication techniques such as frequency shift keying (FSK) or on-off keying (OOK), with wireless communication being the most power-intensive component in the overall system. Wireless communication blocks consume significantly more power compared to other blocks required for functions such as signal detection, digitization, and signal processing. However, far-field data communication offers the advantage of transmitting data over extended distances (several meters or more), depending on antenna size and communication frequency. In far-field data communication, antenna size becomes another challenge, especially in the case of small wireless devices, where this issue is more pronounced [5].

On the other hand, near-field data communication is more suitable for many wireless applications. For instance, near-field data communication is highly advantageous when the distance between devices is only a few centimeters, offering

B. Lee et al., *Energy-management Integrated Circuit Design for Wireless Power and Data Transfer Applications*, Analog Circuits and Signal Processing,
https://doi.org/10.1007/978-3-032-00745-2_4

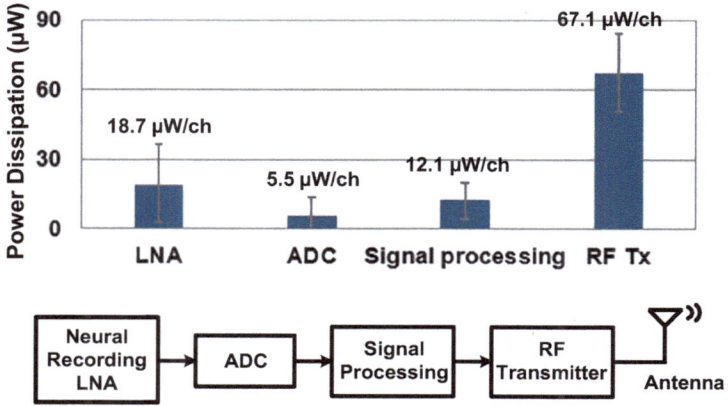

Fig. 4.1 Distribution of power dissipation in the wireless neural recording device [4]

benefits such as lower power consumption and simpler circuit/system design. As a result, near-field data communication is widely employed in scenarios where device power consumption is limited and the system size is small [6].

In recent years, various near-field data communication methods utilizing inductive links and resonant coils have been proposed, which reduce power consumption while enabling wireless data telemetry. The concept of wireless power and data transfer (WPDT), which combines wireless power transfer and wireless data telemetry using inductive links, has been introduced [7, 8]. WPDT is a system that transmits power wirelessly while simultaneously transmitting data, often using the inductive link. Unlike far-field data communication, near-field WPDT eliminates the need for additional components, reducing overall system power consumption and complexity. Near-field data communication is more suitable than traditional far-field data communication for wireless applications where energy efficiency is critical, as it incurs less power loss and has a simpler structure. Such wireless data links can be classified based on transmission speed, distance, uplink/downlink methods, and power consumption. To optimize WPDT systems, proper communication link design is essential, with tailored design considerations based on the operating environment.

4.2 Data Telemetry System Overview

Figure 4.2 presents a simplified block diagram of wireless power and data transmission using an inductive/RF link with two coils in a typical wireless application. This WPDT system provides power through an inductive link composed of a Tx and an Rx LC tank, both tuned to the power carrier frequency (f_p). The data link is divided into a downlink and an uplink, also referred to as forward telemetry and back

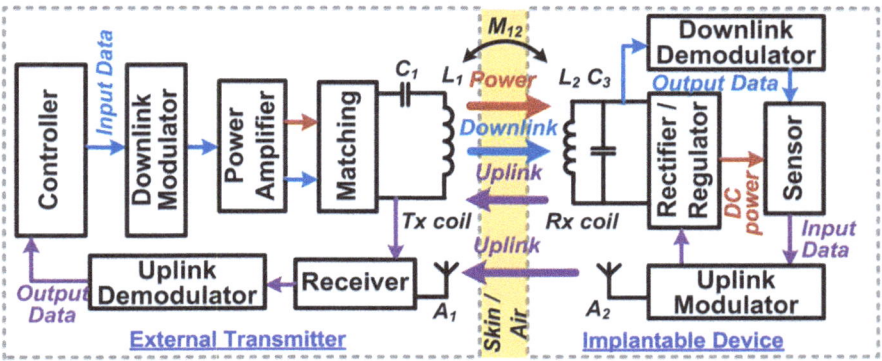

Fig. 4.2 A simplified block diagram of the data and power transfer in the (implantable medical device) IMD through a 2-coil inductive link

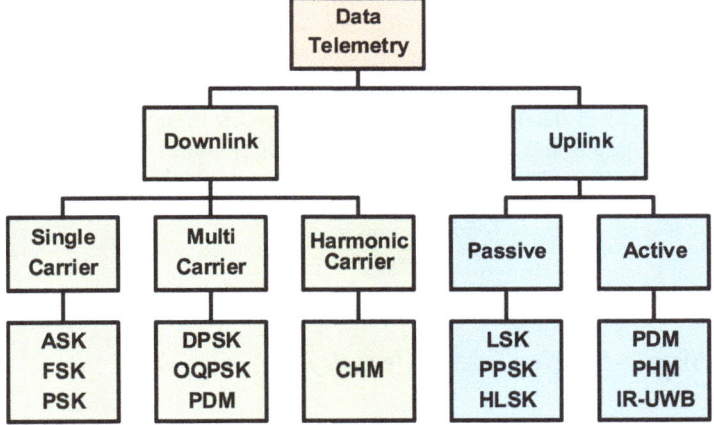

Fig. 4.3 Classification of data telemetry techniques used in wireless applications

telemetry, respectively. The required data transmission speed for the downlink and uplink depends on the application. For example, in systems with a large number of sensors or devices, an uplink speed of several tens of Mb/s may be essential, whereas lower data transmission speeds may suffice for the downlink. Conversely, applications involving real-time device control or processing data from numerous reception channels may require downlink speeds of several tens of Mb/s.

As shown in Fig. 4.3, data telemetry can be broadly classified into downlink and uplink. The downlink can be divided into single-carrier, multi-carrier, and harmonic carrier, while the uplink can be categorized into passive and active modes. Single-carrier telemetry offers simplicity and robustness by modulating the power carrier to transmit data, whereas multi-carrier links separate the power and data carriers, achieving higher PTE and wider bandwidth for both downlink and uplink. In this case, unwanted cross-coupling between the power and data coils must be carefully considered during the design phase to ensure reliable data telemetry by achieving a low bit error rate (BER) and a high signal-to-interference ratio (SIR).

Several traditional modulation techniques for low-power data telemetry have been widely used in single-carrier applications, including amplitude shift keying (ASK) [9], FSK [10, 11], and phase shift keying (PSK) [12, 13]. In contrast, more complex modulation techniques, such as differential PSK (DPSK) [1] and quadrature PSK (QPSK) [13], are employed in multi-carrier telemetry to achieve higher data rates. Recently, a new concept of carrier harmonic modulation (CHM) based on inductive links has been introduced, which offers high data rates without the need for additional complex circuitry or strict timing requirements [14]. In this method, the fundamental frequency of the carrier is used for power delivery, while the harmonics derived from this carrier are utilized for data transmission, optimizing power delivered to the load (PDL) while achieving high data transmission speeds.

In general wireless applications, load shift keying (LSK), a form of passive data telemetry, is still widely used in uplink data telemetry. This is due to its extremely simple structure, potential for miniaturization, and low-power consumption. Recently, pulse-based telemetry methods have been introduced, such as pulse harmonic modulation (PHM) [15], pulse delay modulation (PDM) [16], and impulse radio ultra-wideband (IR-UWB) [17], which allow for high uplink data rates with low-power consumption without the need for a carrier signal. However, these methods may still require additional antennas for data transmission.

4.3 Downlink Data Telemetry

4.3.1 Single-Carrier Data Telemetry

Single-carrier data telemetry utilizes the strong coupling between power coils and the simplicity of coil structure to enable more reliable data transmission even in short distances. Data telemetry using a single carrier is the most widely used modulation method due to its simple modulation and demodulation circuits and low-power consumption. As shown in Fig. 4.4, the most commonly used data modulation techniques in single-carrier data telemetry include ASK, FSK, and PSK, which transmit data by varying their amplitude, frequency, and phase, respectively.

ASK is one of the most widely used modulation methods in downlink data telemetry due to its simple circuit design and low-power consumption [18]. However, there is a trade-off between PTE and data transmission speed. Amplitude-modulated data is sensitive to noise and interference, requiring a higher modulation index. The modulation index refers to the ratio of change in the carrier amplitude between data bits "0" and "1." ASK with a 100% modulation index is known as OOK, which allows for higher data transmission speeds and easier data detection, but results in lower PTE.

Fig. 4.4 Single-carrier
data modulation techniques
for downlink data
telemetry

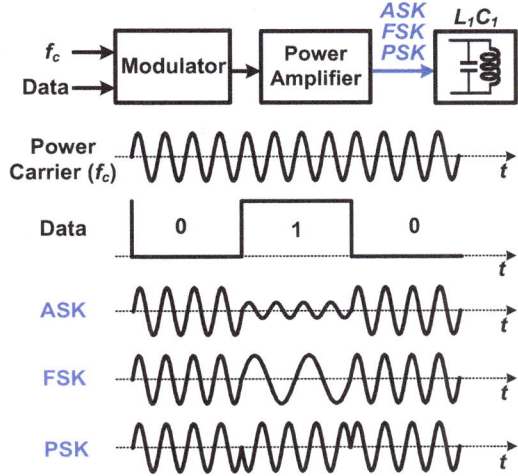

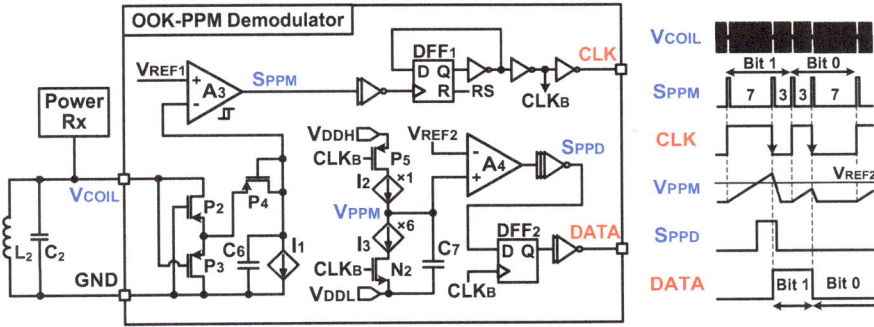

Fig. 4.5 Schematic diagrams of the OOK-PPM demodulator circuit and its waveforms [19]

Additionally, inaccurate synchronization between data and clock signals makes ASK more sensitive to changes in inductive coupling and noise components. To address this, pulse position modulation ASK (PPM-ASK) is used to recover clock-synchronized data along with ASK modulation. Another technique, pulse width modulation ASK (PWM-ASK), improves the signal-to-noise ratio (SNR) by distinguishing between bits "0" and "1" through the duty cycle of the power carrier while maintaining bit-timing information and recovering clock-synchronized data.

Figure 4.5 illustrates the schematic of an OOK-PPM demodulator circuit as a design example [19]. In the OOK demodulator, V_{COIL} is converted to a half-wave through a cross-coupled PMOS pair, P2 and P3, and rectified via a diode-connected transistor, P4. The envelope of V_{COIL} is then fed to a hysteresis comparator, A3, to generate the PPM signal (S_{PPM}). To recover data from the S_{PPM} signal, it first passes through a frequency divider (DFF1) to be converted into the CLK signal. CLK controls the timing and amplitude of V_{PPM}, alternately charging and discharging C_7

through current sources I_2 and I_3, respectively. When the pulse position ratio of S_{PPM} is 7:3, I2 charges C7 for a longer duration, causing V_{PPM} to exceed V_{REF2} when CLK = 1, leading to DATA = 1. Conversely, if the ratio is 3:7, V_{PPM} does not reach V_{REF2}, resulting in DATA = 0. This design uses analog current sources like I_2 and I_3 and a capacitor C_7 to demodulate the data bits. However, for a more robust and low-power design, a fully digital PPM demodulator can also be employed [20].

These methods provide better sensitivity in ASK modulation and reduce power consumption, but compared to conventional ASK, PPM-ASK and PWM-ASK require more time to represent each data bit, resulting in lower data transmission speeds. FSK and PSK are less sensitive to noise, but they require a wider bandwidth per bit than ASK, which can be problematic when using high-Q coils. Moreover, PSK requires synchronization in both frequency and phase, necessitating relatively more complex circuits on both the transmitter and receiver sides.

Figure 4.6a presents the schematic diagrams of a frequency-splitting-based wireless power and FSK data transfer system [21, 22]. This system adopts the FSK method for downlink data transmission, offering a solution to the trade-off between PDL and data rate (DR) observed in the OOK modulation method. FSK enables continuous power supply and data transmission, but in high-Q coils, a trade-off between PTE and DR still persists. Another trade-off in single-link WPDT systems arises from the carrier frequency and bandwidth. As the carrier frequency increases, WPDT systems gain wider bandwidth, which favors higher DR. However, higher carrier frequencies also lead to increased tissue conductivity, resulting in greater tissue absorption and heating. Moreover, power management integrated circuit (PMIC) generally exhibits lower power conversion efficiency (PCE) when operating at higher frequencies. To address these challenges, this system employs FSK data transmission technology in a frequency-splitting link. As shown in Fig. 4.6b, frequency splitting occurs when the distance between coupled coils decreases, causing the peak link frequency (f_{peak}) to split into two frequencies. This system utilizes flat-region FSK (FR-FSK) to resolve the trade-off between PTE and DR. To compensate for the narrow frequency separation, a high-sensitivity FSK demodulator is proposed, supporting high DR and maintaining power transmission performance.

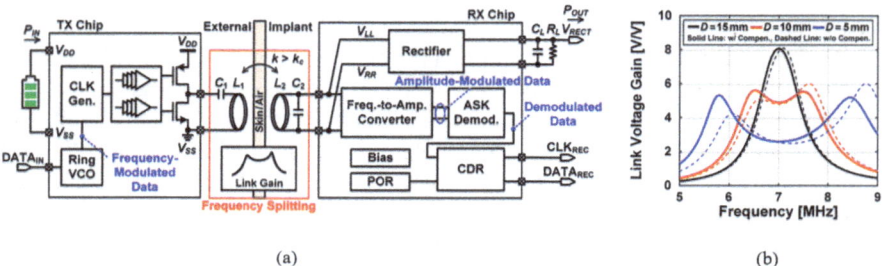

(a)   (b)

Fig. 4.6 (a) Schematic diagrams of the frequency-splitting-based power and FSK data transfer system and (b) link voltage gain varying the carrier frequency for different coil distances [21]

Consequently, the FS-WPDT system using the FR-FSK method achieves large PDL, high DR, and high PTE simultaneously.

Most conventional ASK FT systems require an envelope detector to demodulate the FT data [23] as shown in Fig. 4.7a. When the FT data is "1," the transmitter L_1 voltage (V_{TX}) is set to zero, causing a reduction in the receiver L_2 voltages (V_{IN1} and V_{IN2}). Consequently, the envelope output voltage (V_{ENV}) decreases, and the analog comparator outputs "1" when V_{ENV} drops below the reference voltage (V_{REF}). However, the envelope detector acts as a passive rectifier, drawing power from the L_2C_2 tank, which reduces the PDL. Additionally, the analog comparator used after the envelope detector introduces further power dissipation. The passive RC components required for the envelope detector also occupy a significant silicon area, increasing the chip size in IC designs. Therefore, eliminating the envelope detector can significantly reduce power dissipation and decrease the overall chip size.

The envelope-detector-less (EDL) ASK FT system, as shown in Fig. 4.7b, eliminates the need for an envelope detector. Instead, the proposed method utilizes the outputs of the CG comparator (V_{G1} and V_{G2}), which are primarily used to control the power-pass PMOS transistors. When the FT data becomes "1," V_{IN1} and V_{IN2} decrease below the rectifier output voltage (V_{REG}), causing V_{G1} and V_{G2} to equal V_{REG}. The proposed digital cleaner processes V_{G1} and V_{G2} to demodulate the data into a fully digitized form. In other words, the system achieves fully digital demodulation without relying on an envelope detector. This approach eliminates the additional power

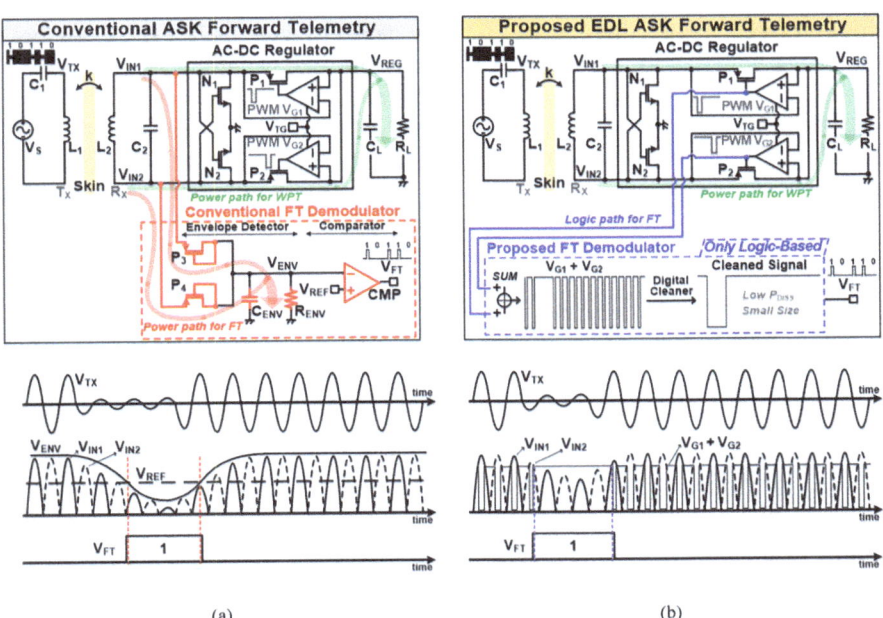

Fig. 4.7 Conceptual diagram of (**a**) conventional ASK FT and (**b**) proposed envelope-detector-less (EDL) ASK FT schemes [23]

path required for data telemetry and removes the analog comparator, resulting in reduced power dissipation. Furthermore, the absence of RC components minimizes the IC size, making the system more efficient in terms of power and silicon area.

In conventional ASK FT systems, the analog components, including the envelope detector and analog comparator, dissipate 123.3 µW of power, along with 30.14 µW consumed by the synchronizer. In contrast, the digital cleaner in the proposed EDL ASK FT system consumes only 64.74 µW, with the synchronizer maintaining the same 30.14 µW power consumption. This results in a power savings of 58.56 µW, corresponding to a 38.2% reduction in total power consumption. Additionally, the proposed EDL ASK FT system was fully implemented in an IC, occupying a compact area of 0.023 mm^2.

Over the past few decades, single-carrier data links have evolved to achieve high-speed data transmission, with developments such as phase-coherent FSK operating at 2.5 Mb/s [24], binary PSK at 1.12 Mb/s [25], and quadrature PSK at 8 Mb/s [13]. However, single-carrier data telemetry has limitations in achieving higher data transmission speeds due to the conflicting requirements between data transmission bandwidth and PTE. This is because the same carrier, which shares the same LC tanks in both the transmitter and receiver, is used.

4.3.2 Multi-Carrier Data Telemetry

To overcome the limitations of single-carrier data telemetry, multi-carrier data links have been proposed to achieve high PTE and wide bandwidth for both uplink and downlink. This approach utilizes multiple pairs of coils or antennas (up to three), each dedicated to a specific function such as power transfer, uplink data, or downlink data transmission. As shown in Fig. 4.8, separating the power carrier from the data carrier is advantageous in high-performance wireless applications requiring wide bandwidth, but this comes at the cost of increased system size and complexity.

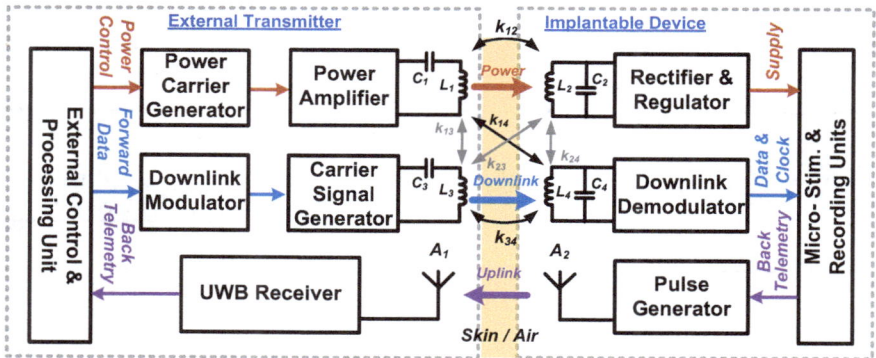

Fig. 4.8 Block diagram of the multi-carrier wireless link and its associated blocks

One of the major challenges in multi-carrier link design is the strong power carrier interference caused by the cross-couplings (k_{14}, k_{13}, k_{23}, k_{24}) between the power and data coils. To address this, it is crucial to add a band-pass filter (BPF) at the Rx to suppress power carrier interference from the received data carrier signal. Alternatively, the shape of the coil pairs can be adjusted to minimize cross-coupling. For instance, using orthogonal coils or figure-8 coils can be effective. In orthogonal data coils, which are symmetrically aligned with the power coils on both the Tx and Rx sides, the magnetic flux coupling between the power and data coils is minimized [26]. In figure-8 coils, the symmetrical but opposite windings of the data coils cancel out the current induced by the power carrier, reducing interference [27].

In [12], asynchronous DPSK was employed to cancel power carrier interference using differential demodulation techniques, achieving a data rate of up to 2 Mb/s without the need for high-order BPFs. Multi-carrier data telemetry was extended to both uplink and downlink in an inductive power transfer system using coplanar coils and offset QPSK, reaching a data rate of 4.15 Mb/s [13]. These multi-carrier techniques play a crucial role in wireless applications by reducing interference between power transfer and data communication while simultaneously achieving high data transmission rates and power efficiency.

4.3.3 Pulse-Based Data Telemetry

Multi-carrier telemetry can achieve higher PTE and data transmission rates compared to single-carrier systems, but securing sufficient bandwidth at high carrier frequencies remains a challenge. To address this, some studies have proposed pulse-based data links, which replace traditional carrier signals with short, sharp pulses [28]. This method simplifies clock recovery at the Rx and improves the SNR, providing enhanced performance. While pulse-based techniques are also employed in non-medical fields, their use in IMDs may be limited due to strong power carrier interference. Low-Q LC tanks offer wide bandwidth but are more susceptible to power carrier interference. Conversely, increasing the Q-factor reduces bandwidth and can lead to inter-symbol interference (ISI).

To overcome these limitations, as shown in Fig. 4.9, a method called PHM has been proposed, which allows high data transmission rates while maintaining a high Q-factor [15]. In PHM, sharp pulses are used to transmit data, with the timing and amplitude of each pulse carefully adjusted to minimize ISI. In this method, a pulse is generated to transmit a bit "1," initiating residual oscillation in the Rx LC tank, while no pulse is transmitted for a bit "0." A second pulse with a specific amplitude and delay is transmitted to suppress residual oscillations, reducing ISI and enabling data rates up to 20 Mb/s. Additionally, the high-Q LC tank provides protection against power carrier interference.

Thanks to its low-power consumption and small chip size, PHM can be used for both uplink and downlink transmissions. When combined with automatic gain control (AGC), it can simplify Rx design. However, this technique is sensitive to the

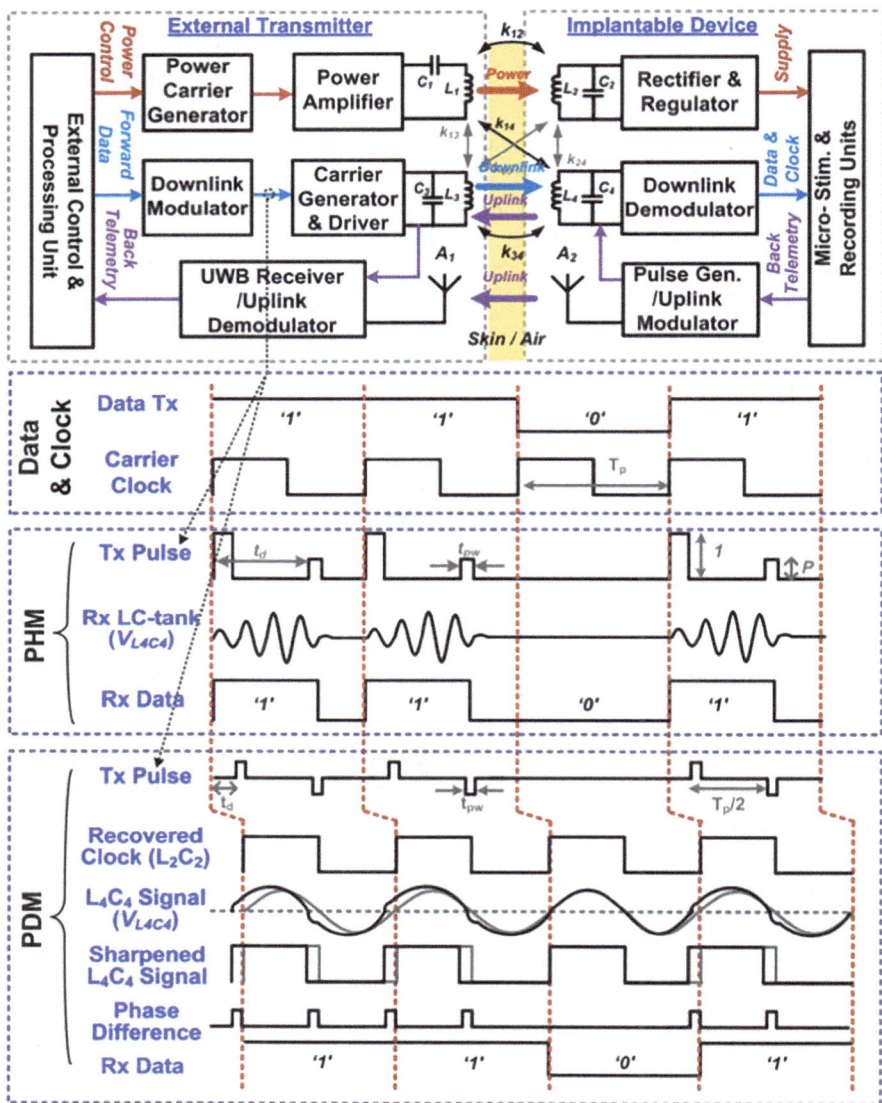

Fig. 4.9 The simplified block diagram and conceptual waveforms for PHM and PDM data transmission methods

delay and amplitude of the pulses, which are influenced by the coupling between the Tx and Rx coils. Moreover, the signal amplitude at the Rx must be greater than the power carrier interference, requiring an SIR >0 dB for proper data reception.

Another modulation method, known as PDM, utilizes power carrier interference to transmit data even with an SIR below 0 dB. In this technique, two narrow pulses are transmitted at a half-cycle interval when the bit is "1," synchronized with the power carrier to avoid interference. No pulses are transmitted for bit "0," but data can still be recovered from the Rx clock. PDM offers robust data transmission even in the presence of strong power carrier interference, making the system simpler and more stable. This approach is particularly suitable for high-power wireless applications. A PDM transceiver demonstrated a data rate of 13.56 Mb/s with power consumption of 162 pJ/bit, achieving performance at an SIR of −18.5 dB.

4.3.4 Harmonic-Based Data Telemetry

Harmonics-based data telemetry introduces a novel approach for WPDT by utilizing the fundamental carrier frequency for power transfer and higher-order harmonics for data transmission [29]. This method eliminates the need for separate data drivers or additional carrier signals, thereby reducing system complexity and power consumption.

Figure 4.10 illustrates the CHM data transmission scheme, wherein specific harmonics (the third and seventh harmonics) are utilized to transmit independent data streams. The fundamental frequency is dedicated to power transfer, ensuring efficient energy delivery, while higher-order harmonics are employed for data transmission, effectively minimizing interference between power and data signals. Furthermore, figure-8-shaped coils are implemented in both the transmitter and receiver to reduce magnetic flux coupling between power and data coils, thereby mitigating cross-interference. The WPDT system functionality was verified using an arbitrary waveform generator in the Tx to produce random data streams. The Tx transmits random data streams through the third and seventh harmonic links at a data rate of 550 kbps each, achieving a total data rate of 1.1 Mbps. The data rate is

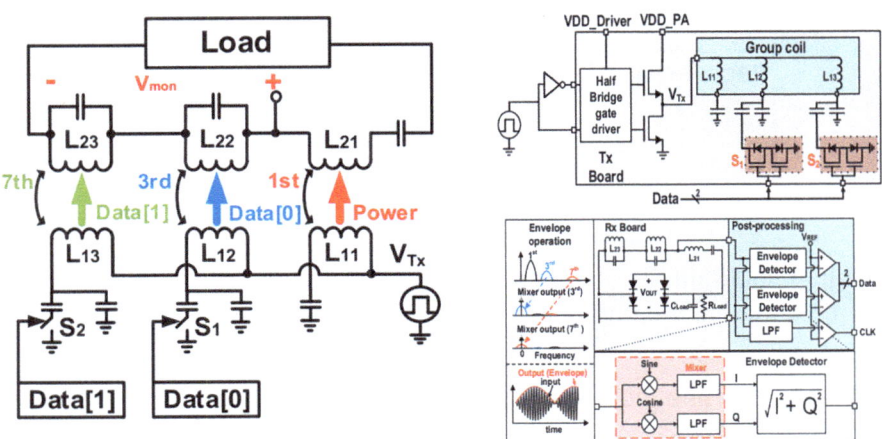

Fig. 4.10 (**a**) The simplified block diagram for CHM methods and (**b**) detailed circuit diagram [29]

designed as a sub-harmonic of the carrier frequency, enabling data recovery at the Rx by utilizing the clock signal extracted from the power carrier.

Figure 4.11 shows the measured waveforms of the CHM data transmission system. The envelope signals of the third and seventh harmonics were consistent with the input individual data streams (Data [0] for the third harmonic, Data [1] for the seventh harmonic). The envelope signals from the third and seventh coils at the Rx were compared with the reference voltage at the frequency of the recovered clock signal. The resulting comparisons were recorded as a 2-bit data stream. The measured BER was $<2.5 \times 10^{-5}$, while the Rx simultaneously received 641 mW of power. Additionally, the power transfer system efficiency was measured to be 17% with a load resistance of 2 Ω. The measured PDL was almost unaffected by the data streams, and the PDL variation was <5% when the load resistance was below 100 Ω, confirming that CHM data transmission minimally impacted power transfer.

In another study of harmonic-based data telemetry [30], a harmonic communication system using a single inductive link was proposed, utilizing the third harmonic for data transmission. This system introduces the WPDT technique that utilized the fundamental and harmonic components of the power amplifier output voltage for efficient power and data transmission, respectively, to mitigate the trade-off between system efficiency and data rate. This system operates at a fundamental frequency of 6.78 MHz for power transfer and used the third harmonic for data transfer. The

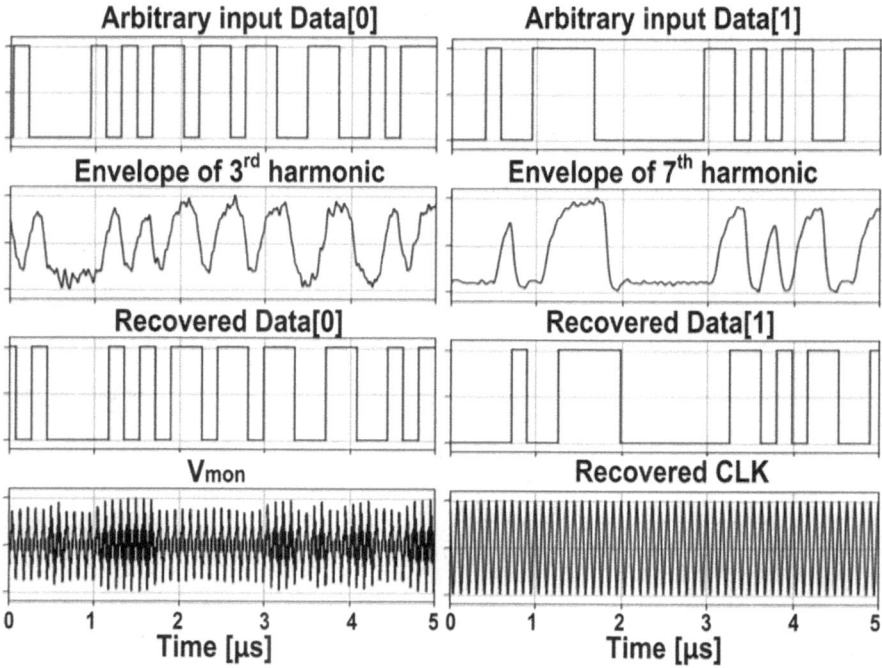

Fig. 4.11 Measured transient waveforms of CHM data transmission with arbitrary input data streams [29]

proposed interference-free rectifier (IFR) is implemented to reduce interference caused by the third harmonic, maintaining a high SNR. Additionally, a tapped coil three capacitor (TL3C) topology was proposed to maximize data channel gain while minimizing its adverse impact on the power channel. This approach achieved a maximum PDL of 82 mW, an end-to-end power transfer efficiency (E2E) of 52.6%, and a data rate of 4 Mbps.

4.4 Uplink Data Telemetry

4.4.1 Passive Uplink Data Telemetry

Generating a carrier signal for data transmission in wireless devices is a significant burden in terms of power consumption, especially in power-constrained applications like biomedical and wearable devices. The data transmission circuit is one of the most power-consuming blocks in biomedical wireless systems. High-performance applications may utilize pulse-based data transfer method such as PHM or PDM. However, due to its simple design and low-power consumption, LSK, a passive modulation scheme, is widely adopted [31]. LSK is commonly found in RFID tags and is well-suited for systems that do not require wide bandwidth.

As shown in Fig. 4.12, LSK method transmits data by modulating the load resistance (R_L) or capacitance (C_2) of the secondary coil (L_2C_2-tank) to modulate the reflected impedance seen by the primary coil. This is achieved by modulating R_L or C_2 according to the serial data bit stream or encoded data. When the coupling coefficient (k_{12}) between the two coils is strong, changes in R_L or C_2 on the secondary side significantly affect the reflected impedance on the primary side. The variation

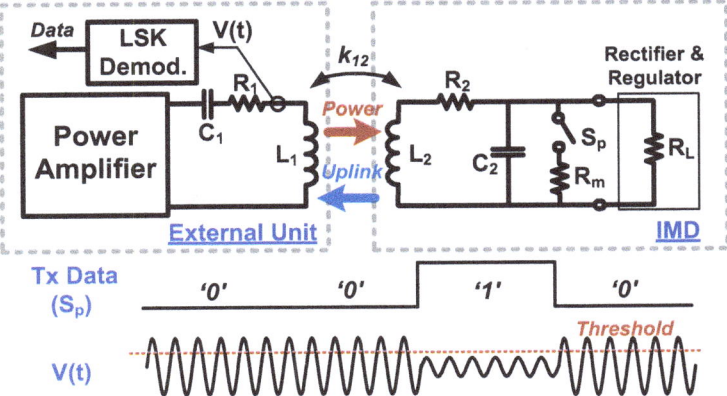

Fig. 4.12 Passive uplink data telemetry by LSK in 2-coil inductive link

in the voltage of the primary coil can be detected to recover the bitstream, operating similarly to ASK demodulation.

However, conventional LSK methods are limited by their effective operation only when the coupling coefficient exceeds a certain threshold. This restriction limits the WPT distance or size of the inductive coils. To address this limitation, various LSK topologies can be implemented in wireless applications to achieve optimal impedance modulation. For low-power consumption devices, where R_L is large, parallel LSK switch is utilized to modulate between R_L and a short-circuit impedance. Conversely, for high-power devices with small R_L, series LSK switches modulate between R_L and an open-circuit impedance. The maximum data rate of LSK method is constrained by the bandwidth of the inductive link, which is approximately proportional to f_P/Q, where f_P is the resonant frequency and Q is the quality factor. In inductive links, the reflected impedance on the primary side is influenced by k_{12}, and the effect of impedance modulation diminishes when k_{12} is low. As such, designing an LSK-based method requires balancing the trade-off between data rate and PTE.

One of the significant drawbacks of LSK is that power cannot be delivered to R_L when the switch is activated. To address this drawback, a passive PSK (PPSK) technique was proposed [32]. PPSK utilizes the transient response of the inductive link to transmit data without significantly impacting power transfer. Figure 4.13 illustrates a simplified block diagram and key waveforms of PPSK when the S_P switch

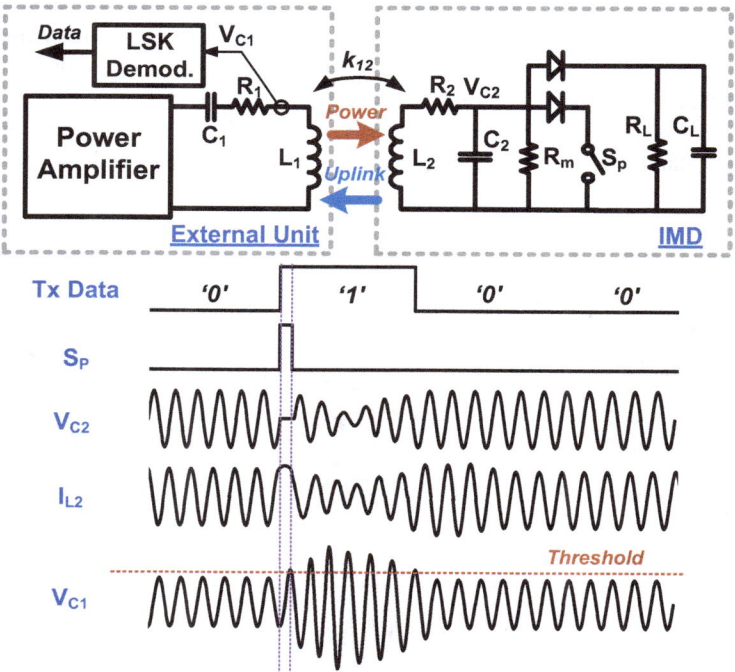

Fig. 4.13 Passive uplink data telemetry by PPSK in 2-coil inductive link

closes to represent a data bit "1." The S_P switch is closed for half a carrier signal cycle, during which the current in the secondary coil (I_{L2}) is maintained at its peak. During this period, the voltage across the secondary coil capacitor (V_{C2}) drops to zero. When the S_P switch opens, I_{L2} realigns with the carrier phase, and the current stored in the secondary coil (L_2) is transferred to the primary coil via the inductive link. This results in an increase in the primary coil current (I_{L1}) and a rapid rise in the primary coil voltage (V_{C1}).

The transient response of PPSK modulation is influenced by the Q-factor of the coil pair and the coupling coefficient (k_{12}), which must be carefully considered during the design process. The PPSK modulation successfully demonstrated uplink data telemetry at a data rate of 1.35 Mbps using a carrier frequency of 13.56 MHz, thereby establishing it as a novel solution for efficient uplink telemetry.

4.4.2 Active Uplink Data Telemetry

For high-performance wireless applications requiring wideband and high data rate uplink telemetry, active data telemetry method becomes essential. For multi-carrier links, uplink data transmission presents significant challenges, primarily due to the substantial power required for data carrier signal generation. Consequently, pulse-based modulation techniques such as IR-UWB, which utilize extremely wide bandwidths of 3–10 GHz, have been widely adopted in wireless systems. IR-UWB transmitters are characterized by low complexity and power consumption, as they can be easily implemented to generate sharp digital pulses for wideband antennas. However, IR-UWB receivers may demand higher complexity and power consumption to recover the transmitted data.

Recently, IR-UWB has been extensively utilized in wireless applications demanding uplink data rates ranging from tens to hundreds of megabits per second with low-power consumption. In UWB telemetry, PPM and OOK techniques are commonly used alongside energy detection. While UWB-OOK transmits a train of impulses only for data bit "1," UWB-PPM generates two time-shifted pulses to represent bits "1" and "0" [33]. Although OOK necessitates threshold estimation, UWB-PPM requires only energy comparisons at two distinct time intervals, which renders it more robust for demodulating the received data bitstream and thus achieving a better BER.

A novel concept of IR-UWB, pulse width modulation IR-UWB (PWM-IR-UWB), has been proposed, as discussed in [34]. This approach achieves higher data rates than PPM while reducing power consumption at the transmitter side. Figure 4.14 illustrates a simplified block diagram and key waveforms of PWM-IR-UWB compared to conventional digital communication using UWB-OOK for N-bit data packets. The PWM-IR-UWB technique employs an analog-to-time converter (ATC) to generate a PWM duty cycle corresponding to the N-bit information. This information is transmitted across the link using two consecutive pulses, eliminating the need for a high-speed analog-to-digital converter (ADC) at the transmitter.

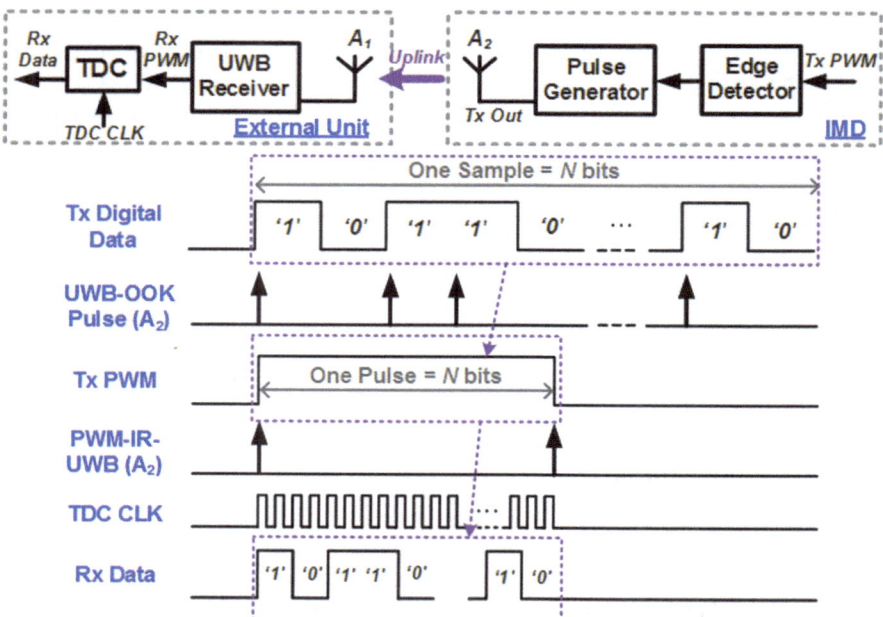

Fig. 4.14 The concept of PWM-IR-UWB data telemetry, which can transmit the equivalent of N digital data bits using only two pulses, compared to the conventional UWB-OOK digital data communication

To convert the received PWM pulses into digital information, a time-to-digital converter (TDC) block is required. However, achieving high resolution with this block typically demands a high-speed clock or minimal gate delays, resulting in high-power consumption. This can consequently impose a significant power burden on power-constrained devices. Therefore, this method is more suitable for uplink scenarios where a high-performance TDC is implemented in external devices, such as high-speed field-programmable gate arrays (FPGAs). This allows for accurate measurement of the time delay between successively received PWM-IR-UWB pulses and subsequent recovery of the N-bit data. This technique can achieve high data rates of up to 100 Mbps, but its reliability depends on the wireless link bandwidth and the accuracy of the ATC and TDC.

4.4.3 Advanced Uplink Data Telemetry

The design of wireless telemetry systems often faces a fundamental trade-off between power transfer efficiency and data rate. This challenge is particularly critical in applications with power and size constraints, such as medical implants, wireless sensors, and IoT devices. To address this challenge, the proposed a novel telemetry IC that enables simultaneous power transfer and high-speed data communication using cyclic on-off-keying (COOK) modulation over a single 13.56 MHz inductive link [35].

As shown in Fig. 4.15, the COOK modulation method differs from conventional LSK by adopting a different approach. COOK method minimizes energy losses caused by uplink data transmission while maintaining high resonant quality, by synchronizing short-cycle shorts with the resonant conditions of the LC tank. This allows for efficient data encoding and transmission while effectively maintaining the power transfer capability of the inductive link. The COOK-based telemetry system achieved a high data rate of 6.78 Mbps at a carrier frequency of 13.56 MHz. The IC power consumes only 64 μW, resulting in an energy consumption of 9.5 pJ per bit, thereby demonstrating exceptional efficiency.

In conventional LSK data transmission, the configuration of the LSK switch (serial or parallel) is adjusted according to the power consumption level of the receiver to achieve optimal impedance modulation. However, conventional short-circuit LSK (SC-LSK) and open-circuit LSK (OC-LSK) methods face significant challenges in stable data transmission under varying load conditions. For instance, SC-LSK is optimized for light load conditions, whereas OC-LSK is optimized for heavy load conditions. As a result, SC-LSK is unsuitable when the full channel of recordings and stimulations is activated, and OC-LSK cannot be used during the sleep mode of the applications. This limitation occurs because the magnitude of the reflected impedance change on the primary side (Z_{IN}) varies significantly with load conditions, causing the primary coil voltage change (ΔV_{L1}) to depend on Z_{IN}. To address this issue, [36] proposed a hybrid-LSK uplink data telemetry system, which enables load-insensitive uplink data transmission in receivers powered by inductive links.

By combining SC and OC operations, hybrid-LSK maintains a consistent ΔV_{L1} regardless of load variations. Intuitively, ΔV_{L1} is defined as the difference between the maximum V_{L1} (in SC-LSK mode) and the minimum V_{L1} (in OC-LSK mode). Analytically, the ΔV_{L1} function does not depend on Z_{IN}, whereas SC-LSK and OC-LSK functions do. Therefore, hybrid-LSK can be implemented across a wide

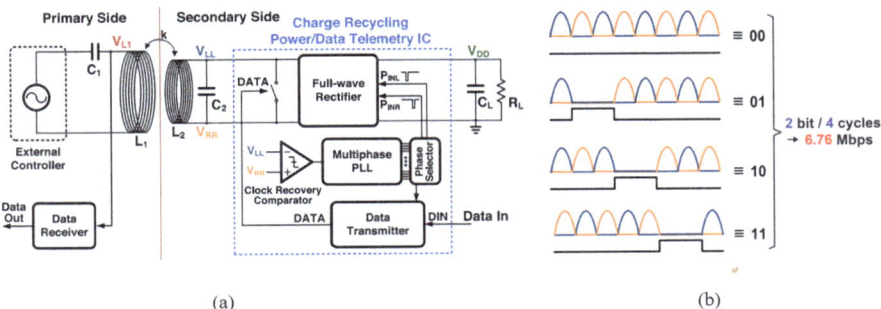

(a)   (b)

Fig. 4.15 (**a**) Block diagram of the implemented system for power and data telemetry and COOK modulation scheme and (**b**) its conceptual waveforms [35]

range of load conditions, making it suitable for IMDs or IoT devices operating in both active and sleep modes. Furthermore, by using a different sequence of SC and OC operations, two hybrid-LSK types can be designed: hybrid-LSK 1 (OC → SC) and hybrid-LSK 2 (SC → OC). Hybrid-LSK 1 is inefficient because the energy stored in the L_2C_2 tank is wasted during the SC duration, despite its shorter OC duration. On the other hand, hybrid-LSK 2 transfers the built-in energy from the L_2C_2 tank to the load but suffers from a longer OC duration. As a result, there is a trade-off between energy efficiency and OC duration, which can be optimized based on the choice of designer. Moreover, ΔV_{L1} is predictable both through simulation and analytical modeling. With fixed duration of SC or OC operation and calculated ΔV_{L1}, the slope of ΔV_{L1} is predictable and remains consistent across all load conditions. This predictability enables the design of a slope-based demodulator, which demodulates the ΔV_{L1} signal generated by hybrid-LSK into digitized data over a wide range of load conditions.

The conventional SC-LSK exhibits a ΔV_{L1} variation of 510 mV between its maximum and minimum values across load conditions ranging from 50 Ω to 50 kΩ, whereas the proposed hybrid-LSK 2 achieves a significantly reduced variation of 60 mV. This corresponds to 450 mV reduction in ΔV_{L1} variation and a 34.1% reduction in VAR$_{VC}$, defined as the difference between the maximum and minimum ΔV_{L1} normalized to the maximum ΔV_{L1}. Additionally, the proposed hybrid-LSK system achieves a BER below 9.1×10^{-10} across a wide range of load conditions (50 Ω to 50 kΩ), while conventional LSK systems maintain low BER only under specific load conditions (Fig. 4.16).

4.5   Bidirectional Data Telemetry

Bidirectional data telemetry is a core technology in wireless applications, enabling data exchange between Tx and Rx [37]. It plays a particularly critical role in applications such as IoT, medical devices, wearable sensors, and portable devices. Through bidirectional data telemetry, Tx and Rx can communicate by exchanging data. Rx collects recorded data and sends it to Tx, while Tx processes the data to control Rx's operation or adjust control parameters. For example, within biomedical systems, the Rx monitors a patient's physiological signals and transmits the recorded data to Tx, which can send data back to Rx to adjust stimulation parameters as needed. Monitoring Rx device status and feedback controls are essential for maintaining the stability of wireless systems. Rx can transmit device status or recorded data to Tx to detect potential errors in advance, and Tx can respond with immediate control signals to ensure the stability of the wireless system.

As shown in Fig. 4.17, bidirectional data telemetry is categorized into half-duplex (HD) and full-duplex (FD) based on the data telemetry method. As shown in Fig. 4.17a, in HD systems, data can only be transmitted in one direction at a time. For instance, when Tx sends downlink data to Rx, Rx cannot simultaneously send uplink data to Tx. HD systems alternate between data transmission directions,

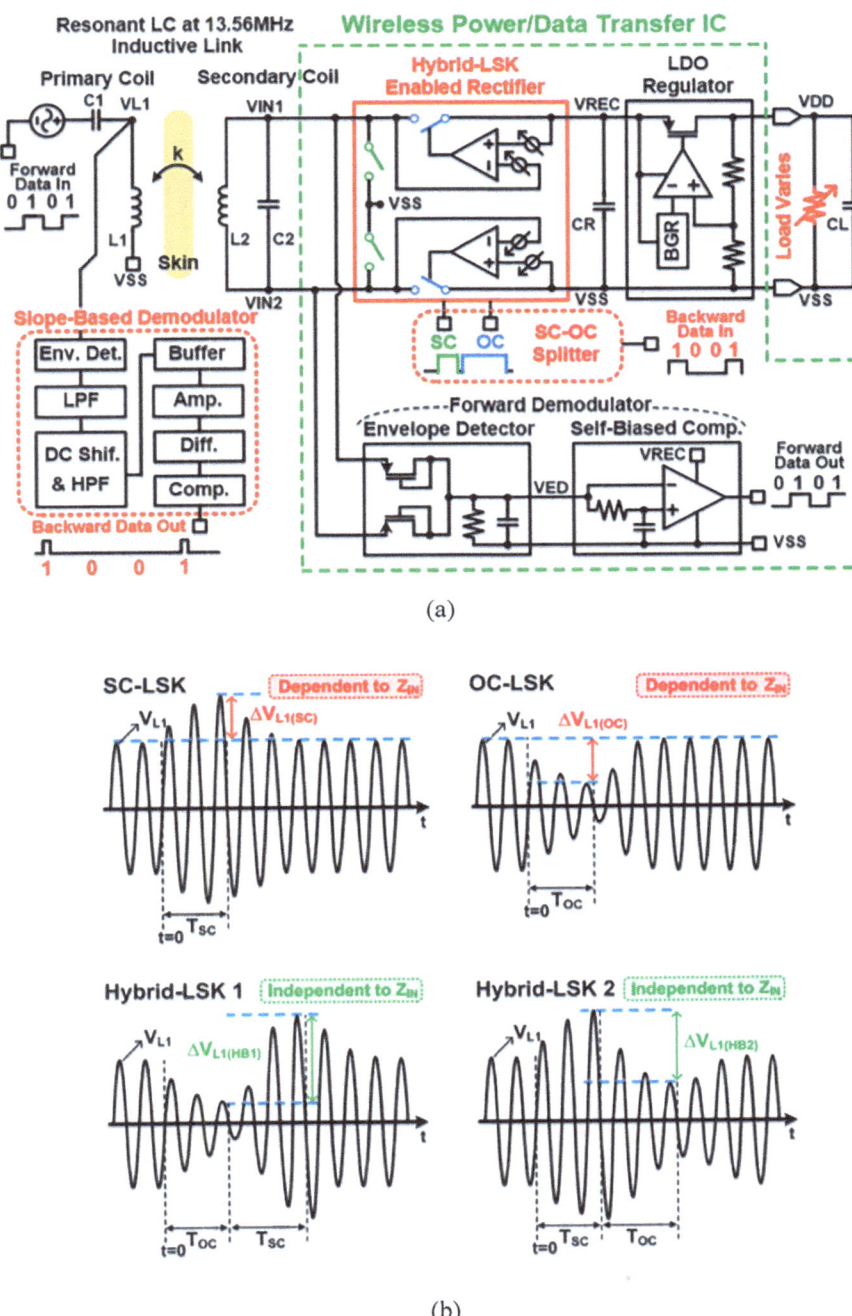

(a)

(b)

Fig. 4.16 (**a**) Overall structure of the wireless power and data transfer IC with the hybrid-LSK uplink data telemetry and (**b**) its conceptual waveforms of hybrid-LSK scheme [36]

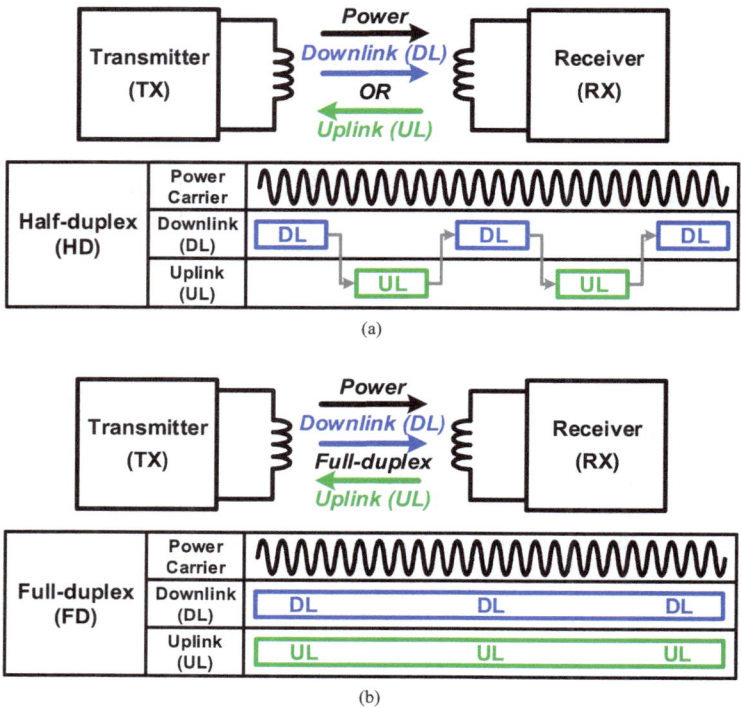

Fig. 4.17 Illustration and time sequence of WPDT with (**a**) HD and (**b**) FD methods [37]

separating the time intervals for downlink and uplink data transmissions. Conversely, as shown in Fig. 4.17b, FD systems enable simultaneous bidirectional data telemetry. Downlink and uplink data are exchanged independently, either through bandwidth or frequency separation, or via advanced signal processing techniques to mitigate interference. The choice between HD and FD systems depends on the specific application, and careful design is required to meet the requirements of the target system.

4.5.1 Half-Duplex (HD) Data Telemetry

The half-duplex (HD) method is a system where the transmitter and receiver share the same data channel and can only transmit data in one direction at a time. Since the direction of data transmission alternates, downlink and uplink data cannot be transmitted simultaneously. This characteristic facilitates the HD method used in various wireless applications, particularly in those requiring structural simplicity and energy efficiency.

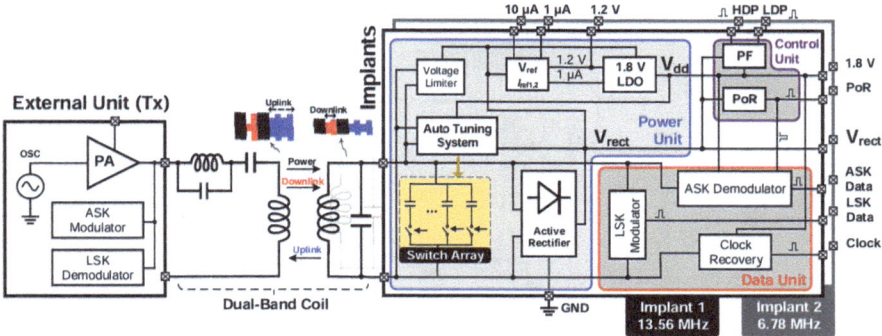

Fig. 4.18 Block diagram of the wireless power and bidirectional data communication system with global power control for the closed-loop biomedical implants [38]

As shown in Fig. 4.18, the WPDT system using HD method is demonstrated for multi-site IMD applications [38]. This system is designed to efficiently transmit both power and data through a single inductive link. It employs adaptive resonance tuning to dynamically adjust the resonant frequency of the inductive link in real time, optimizing the PTE. This system utilizes ASK modulation for downlink and LSK modulation for uplink to transmit bidirectional data. By alternating between downlink and uplink data telemetry in HD, bidirectional data is effectively transmitted through a single shared channel.

Additionally, the transmitter utilizes dual-band coils operating at 13.56 MHz and 6.78 MHz to independently control multiple implants. This system utilizes frequency division multiple access (FDMA) to minimize interference between implants and ensures the independent transmission of data and power to each. Since the system transmits power and data simultaneously, it incorporates a closed-loop power feedback system that continuously monitors overvoltage or undervoltage conditions within the implants, thereby ensuring stable power delivery.

Figure 4.19 illustrates a wireless battery charging system that utilizes a single inductive link to charge a battery while simultaneously enabling bidirectional data communication [39]. This system employs the HD method to integrate both power and data transmission, specifically designed to enhance the safety and efficiency of power delivery in IMDs. This system implements bidirectional telemetry using binary PSK (BPSK) modulation for downlink and pulsed-LSK modulation for uplink. The wireless battery charging system uses downlink data telemetry to transmit system status information, such as power delivery status or battery charging requirements, and to control the receiver's operation. Conversely, uplink data telemetry provides feedback on the receiver's battery voltage status and charging process, allowing the transmitter to adjust power output and prevent unnecessary power loss during charging, thereby improving overall system efficiency. This bidirectional data telemetry continuously monitors the battery charging state, efficiently integrating power transfer and data transmission within the wireless charging process.

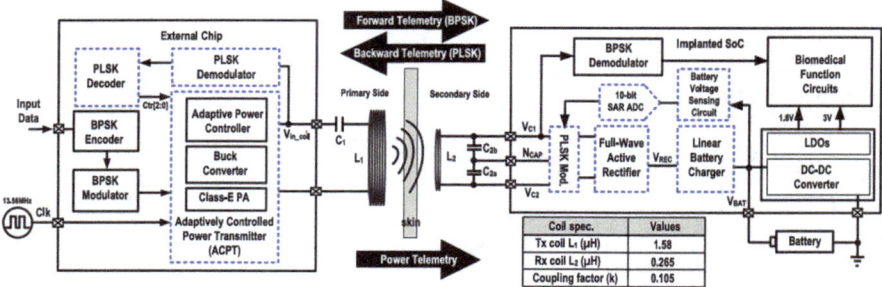

Fig. 4.19 Block diagram of the proposed wireless battery charging system with global power control which is integrated with an implantable closed-loop epileptic seizure control SoC [39]

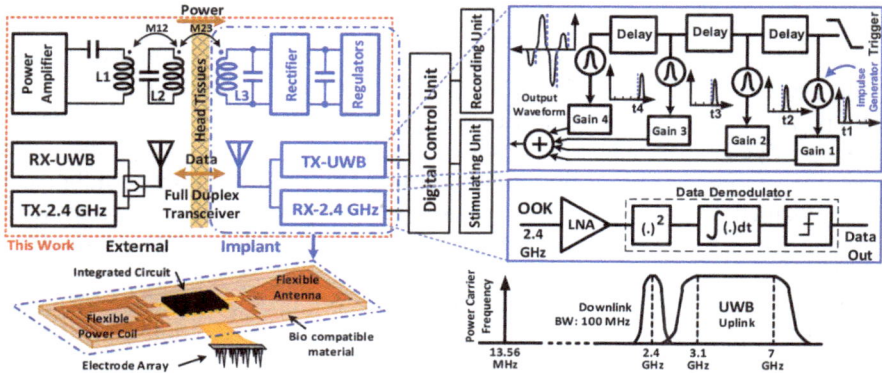

Fig. 4.20 Block diagram of an implantable neural recording and stimulating device including an inductive power link and the proposed full-duplex transceiver [40]

4.5.2 Full-Duplex (FD) Data Telemetry

FD systems enable the simultaneous transmission and reception of downlink and uplink data. Due to their capability to transmit bidirectional data concurrently, FD systems achieve higher data rates and facilitate real-time control between the transmitter and receiver. However, FD systems require additional circuits or algorithms for self-interference cancelation, ensuring that the transmitted signal does not adversely affect the received signal.

Figure 4.20 illustrates a system designed to implement high-speed FD data telemetry for IMD applications such as neural stimulation and recording [40]. The proposed system utilizes a single RF antenna to simultaneously process both uplink data (neural signal information) and downlink data (stimulation parameters). The use of a single, compact antenna contributes to the miniaturization of IMDs and reduces the hardware complexity of the device.

In FD systems, simultaneous downlink and uplink transmission using a single antenna may lead to interference where the downlink signal affects the receiver. To mitigate this issue, self-interference cancelation technology was applied. The system was designed to minimize interference by operating downlink telemetry in the 2.4 GHz industrial, scientific, and medical (ISM) band and uplink telemetry in the 3.1–7 GHz UWB band, thereby separating the two across different frequency ranges. Additionally, low-noise amplifier (LNA) and high-frequency filters were employed to suppress uplink transmission signals while accurately recovering downlink signals.

This system successfully implements real-time bidirectional data telemetry, achieving a maximum downlink data rate of 100 Mbps and a maximum uplink data rate of 500 Mbps. By utilizing FD technology with a single antenna, the system simultaneously processes downlink data, such as stimulation commands, and uplink data, such as neural recording information, thereby enabling a fully closed-loop system.

Figure 4.21 illustrates an FD data telemetry system for underwater wireless sensor network (UWSN), which simultaneously handles wireless power transfer and bidirectional data communication [41]. This system is implemented to concurrently process downlink and uplink data over a single inductive link. For downlink data telemetry, the system utilizes ASK modulation method, integrating downlink data transmission with power delivery by combining the data with the power transfer signal. For uplink data telemetry, the system employs OOK modulation, wherein the receiver transmits data to the transmitter by modulating the load variations on the power transfer signal. To mitigate self-interference between downlink and

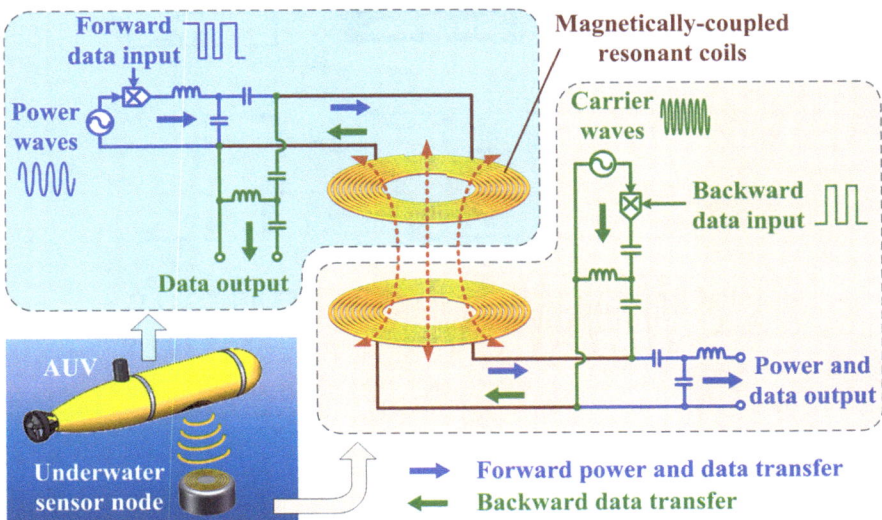

Fig. 4.21 Block diagram of a novel simultaneous wireless power and data system is proposed for underwater wireless sensor networks [41]

uplink channels in this FD system, the double-LCC and double-CLC networks are employed in the Tx and Rx, respectively. These networks facilitate the independent design of the impedance for the power and data channels, thereby effectively minimizing cross-channel interference. This system achieves a downlink data rate of 20 kbps and an uplink data rate of 300 kbps, while delivering up to 23.1 W of power, thus ensuring stable power delivery even during data transmission.

The previous systems implemented FD data telemetry by minimizing interference between the power and data channels through the use of antennas or additional capacitors and inductors as couplers. Figure 4.22 illustrates a system that simultaneously transmits power and FD data without requiring separate antennas or couplers [42]. This system achieves FD data telemetry and power transfer through a 2-coil inductive link eliminating the need for additional coils or antennas. The downlink utilizes dual-band FSK modulation, enabling efficient power delivery and data transmission across two resonant frequencies. For the uplink, LSK modulation is employed, transmitting data through impedance variations reflected to the Tx. The system employs a dual-band coil design to optimize resonant frequencies and impedance at two frequencies, thereby minimizing interference between the

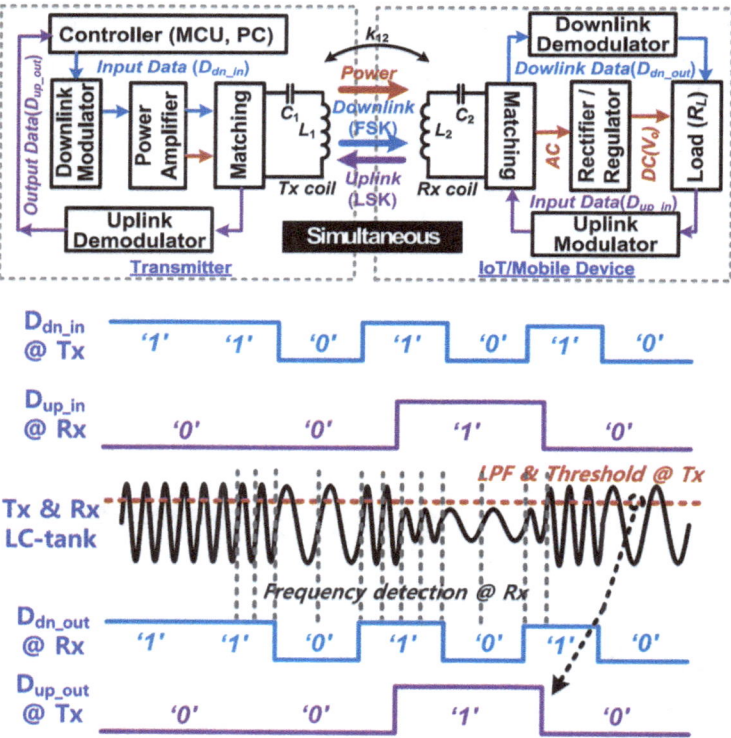

Fig. 4.22 Block diagram of the power and FD data telemetry via a 2-coil inductive link without additional components [42]

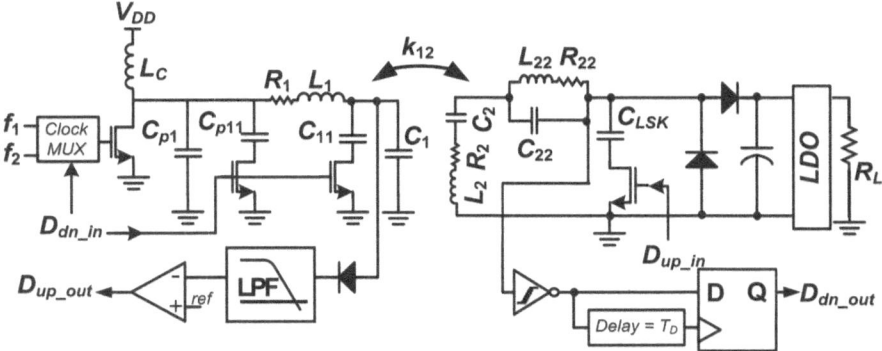

Fig. 4.23 Schematic of the power and FD data telemetry via a 2-coil inductive link [42]

downlink and uplink. The FSK downlink demonstrates robustness against self-interference caused by LSK, enabling reliable FD data telemetry and power transfer.

The Tx and Rx in this system are designed with dual-band coils that support two resonant frequencies: 2 MHz and 4 MHz. The dual-band coil in the Rx comprises an L_2C_2 tank and the $L_{22}C_{22}$ branch, enabling two resonant frequencies, f_1 and f_2. L_{22} and C_{22} represent lumped circuit elements, specifically a high-quality inductor and capacitor, respectively. The $L_{22}C_{22}$ tank exhibits inductive behavior when the operating frequency aligns with f_1 (the lower frequency) and capacitive behavior when the operating frequency aligns with f_2 (the higher frequency). This implies that the $L_{22}C_{22}$ tank can be effectively modeled as either an inductor (L_{Z22}) or a capacitor (C_{Z22}) outside its resonant frequencies. As shown in Fig. 4.23, the series combination of the $L_{22}C_{22}$ tank and the L_2C_2 tank, forming the dual-band coil module, enables the achievement of two distinct resonant frequencies. The dual-band coil is optimized for high-quality factor values, based on the dual-band coil analysis presented in [43]. In WPT applications, the Tx typically faces fewer physical constraints compared to the Rx, which facilitates the straightforward implementation for the dual-band module with additional capacitors (C_1 and C_{p11}) and switches. When the input downlink data, D_{dn_in}, is "1," C_1 and C_{p11} are bypassed, and the power carrier shifts from the lower resonant frequency to the higher resonant frequency. The class-E power amplifier in the Tx, integrated with a dual-band FSK data modulator, is optimized for the system's two resonant frequencies and high efficiency, thereby ensuring a stable envelope for the power carrier.

As shown in Fig. 4.24, the proposed system achieved a downlink data rate of up to 800 kbps, an uplink data rate of up to 14 kbps, and a maximum power transfer of 140 mW. The BER for the downlink and uplink were measured at 3.12×10^{-5} and below 10^{-6}, respectively, demonstrating high reliability of data transmission. By employing a dual-band design, the system facilitated stable power delivery and FD data telemetry without the need for additional components. This approach offers a novel solution for applications such as IoT and IMDs, where real-time data transmission and efficient power management are essential for compact devices.

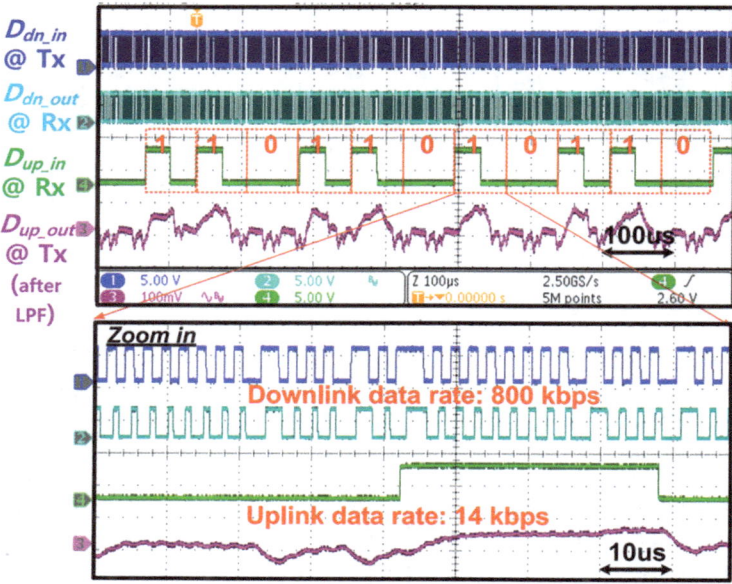

Fig. 4.24 Measured transient waveforms for the power and FD data telemetry via a 2-coil inductive link [42]

4.6   Conclusions

This chapter has categorized and explained wireless data telemetry for wireless applications, including the IoT and wearable biomedical devices, based on uplink, downlink, and bidirectional data telemetry. Given that even state-of-the-art data telemetry systems cannot yet fully address all considerations for remote monitoring of physiological signals, designers must meticulously select the most appropriate data telemetry system while striving for optimal performance. This involves taking into account the practical limitations of the application and the available resources, particularly concerning power consumption, physical area, data rate, reliability, communication distance, and other relevant factors.

References

1. Y.K. Lo, K. Chen, P. Gad, W. Liu, A fully-integrated high-compliance voltage SoC for epi-retinal and neural prostheses. IEEE Trans. Biomed. Circuits Syst. **7**(6), 761–772 (2013)
2. A. Borna, K. Najafi, A low power light weight wireless multichannel microsystem for reliable neural recording. IEEE J. Solid State Circuits **48**(2), 439–451 (2014)
3. S. Lee, B. Lee, M. Kiani, M. Ghovanloo, An inductively-powered wireless neural recording system with a charge sampling analog front-end. IEEE Sensors J. **16**(2), 475–484 (2016)

4. K. Abdelhalim, L. Kokarovtseva, J.L.P. Velazquiez, R. Genov, 915-MHz FSK/OOK wireless neural recording SoC with 64 mixed-signal FIR filters. IEEE J. Solid State Circuits **48**(10), 2478–2493 (2013)

5. B. Lee, M. Ghovanloo, An overview of data telemetry in inductively powered implantable biomedical devices design and implementation of devices. IEEE Commun. Mag. **57**(2), 74–80 (2019)

6. H.-J. Kim, H. Hirayama, S. Kim, K.J. Han, R. Zhang, J.-W. Choi, Review of near-field wireless power and communication for biomedical applications. IEEE Access **5**, 21264–21285 (2017)

7. S. Rao, N. Liombart, Miniature implantable and wearable on-body antennas: towards the new era of wireless body-centric systems. IEEE Antennas Propag. Mag. **56**(1), 271 (2014)

8. B. Lee, M. Kiani, M. Ghovanloo, A triple-loop inductive power transmission system for biomedical applications. IEEE Trans. Biomed. Circuits Syst. **10**(1), 138–148 (2015)

9. Y. Chen, Y. Liu, Y. Li, G. Wang, M. Chen, An energy-efficient ASK demodulator robust to power-carrier-interference for inductive power and data telemetry. IEEE Trans. Biomed. Circuits Syst. **16**(1), 108–118 (2022)

10. P.R. Troyk, G.A. DeMichele, Inductively-coupled power and data link for neural prostheses using a class-E oscillator and FSK modulation, in *Proc. IEEE 25th EMBS Conf.*, (IEEE, 2003), pp. 3376–3379

11. M.M. Ahmadi, S. Ghandi, A class-E power amplifier with wideband FSK modulation for inductive power and data transmission to medical implants. IEEE Sensors J. **18**(17), 7242–7252 (2018)

12. M. Zhou, M.R. Yuce, W. Liu, A non-coherent DPSK data receiver with interference cancellation for dual-band transcutaneous telemetries. IEEE J. Solid State Circuits **43**, 2203–2012 (2008)

13. G. Simard, M. Sawan, D. Massicotte, High-speed OQPSK and efficient power transfer through inductive link for biomedical implants. IEEE Trans. Biomed. Circuits Syst. **4**(3), 192–200 (2010)

14. Z. Yan, C. Ma, Z. Zhang, L. Huang, P.A. Hu, Redefining the channel bandwidth for simultaneous wireless power and information transfer. IEEE Trans. Ind. Electron. **69**(7), 6881–6891 (2022)

15. F. Inanlou, M. Kiani, M. Ghovanloo, A 10.2 Mbps pulse harmonic modulation based transceiver for implantable medical devices. IEEE J. Solid State Circuits **46**(6), 1296–1306 (2011)

16. M. Kiani, M. Ghovanloo, A 13.56-Mbps pulse delay modulation based transceiver for simultaneous near-field data and power transmission. IEEE Trans. Biomed. Circuits Syst. **9**(1), 1–11 (2014)

17. Y. Gao, Y. Zheng, S. Diao, W.D. Toh, C.W. Ang, M. Je, C.H. Heng, Low-power ultrawideband wireless telemetry transceiver for medical sensor application. IEEE Trans. Biomed. Eng. **58**(3), 768–772 (2011)

18. T.J. Lee, C.L. Lee, Y.J. Ciou, C.C. Huang, C.C. Wang, All-MOS ASK demodulator for low-frequency applications. IEEE Trans. Circuits Syst. II, Express Briefs **55**(5), 474–478 (2008)

19. H.-M. Lee, K.Y. Kwon, W. Li, B. Howel, W.M. Grill, M. Ghovanloo, A power-efficient switched-capacitor stimulating system for electrical/optical deep-brain stimulation. IEEE J. Solid State Circuits **50**(1), 360–374 (2015)

20. H.-M. Lee, C.S. Juvekar, J. Kwong, A.P. Chandrakasan, A nonvolatile flip-flop-enabled cryptographic wireless authentication tag with per-query key update and power-glitch attack countermeasures. IEEE J. Solid State Circuits **52**(1), 272–283 (2017)

21. Y. Park et al., A wireless power and data transfer IC for neural prostheses using a single inductive link with frequency-splitting characteristic. IEEE Trans. Biomed. Circuits Syst. **15**(6), 1306–1319 (2021)

22. Y. Park, P.D. Hung, D. Youn, D. Kwon, C. Kim, M. Je, An enhanced-frequency-splitting-based wireless power and data transfer system achieving 60.2% end-to-end efficiency and 1 Mb/s data rate with a sub-cm RX coil for miniaturized implants, in *2025 ISSCC, San Francisco, CA, USA*, (IEEE, 2025), pp. 1–3

23. H.-S. Lee, H.-M. Lee, A power-efficient envelope-detector-less amplitude-shift-keying forward telemetry for wirelessly powered biomedical devices. IEEE Trans. Biomed. Circuits Syst. **19**(2), 374–384 (2025)

24. M. Ghovanloo, K. Najafi, High data rate frequency shift keying demodulation for wireless biomedical implants. IEEE Trans. Circuits Syst. I, Regul. Papers **51**(12), 2374–2383 (2004)
25. Y. Hu, M. Sawan, A fully integrated low-power BPSK demodulator for implantable medical devices. IEEE Trans. Cir. Syst. I, Regul. Papers **52**(12), 2552–2562 (2005)
26. U. Jow, M. Ghovanloo, Optimization of data coils in a multiband wireless link for neuroprosthetic implantable devices. IEEE Trans. Biomed. Circuits Syst. **4**(5), 301–310 (2010)
27. U. Jow, M. Ghovanloo, Modeling and optimization of printed spiral coils in air, saline, and muscle tissue environments. IEEE Trans. Biomed. Circuits Syst. **3**(5), 339–347 (2009)
28. M. Kiani, M. Ghovanloo, Centimeter-range inductive radios, in *Ultra-Low-Power Short-Range Radios: Centimeter-Range Inductive Radios*, ed. by P.P. Mercier, A.P. Chandrakasan, (Springer International, 2015)
29. J. Lim, B. Lee, Carrier harmonic modulation for simultaneous wireless information and power transfer. IEEE Access **11**, 135568–135574 (2023)
30. Q. Zhuang et al., A 6.78-MHz wireless power and data transfer system achieving simultaneous 48.6% end-to-end efficiency and 4.0-Mb/s forward data delivery with interference-free rectifier. IEEE J. Solid-State Circuits (2025). https://doi.org/10.1109/JSSC.2025.3541290
31. Y.P. Lin, K.T. Tang, An inductive power and data telemetry subsystem with fast transient low dropout regulator for biomedical implants. IEEE Trans. Biomed. Circuits Syst. **10**(2), 435–444 (2016)
32. D. Jiang et al., An integrated passive phase-shift keying modulator for biomedical implants with power telemetry over a single inductive link. IEEE Trans. Biomed. Circuits Syst. **11**(1), 64–77 (2017)
33. M.S. Chae et al., A 128-channel 6 mW wireless neural recording IC with spike feature extraction and UWB transmitter. IEEE Trans. Neural Syst. Rehabil. Eng. **17**(4), 312–321 (2009)
34. J. Lim et al., An impulse radio PWM-based wireless data acquisition sensor interface. IEEE Sensors J. **19**(2), 603–614 (2018)
35. S. Ha et al., Energy recycling telemetry IC with simultaneous 11.5 mW power and 6.78 Mb/s backward data delivery over a single 13.56 MHz inductive link. IEEE J. Solid State Circuits **51**(11), 2664–2678 (2016)
36. H.-S. Lee, J. Ahn, M. Kang, H.-M. Lee, A load-insensitive hybrid LSK back telemetry system with slope-based demodulation for inductively powered biomedical devices. IEEE Trans. Biomed. Circuits Syst. **16**(4), 651–663 (2022)
37. J. Lee, Y. Kim, D. Kang, I. Song, B. Lee, A reconfigurable bidirectional wireless power and full-duplex data transceiver IC for wearable biomedical applications. IEEE Trans. Biomed. Circuits Syst. (2024). https://doi.org/10.1109/TBCAS.2024.3483950
38. M.J. Karimi, M. Jin, Y. Zhou, C. Dehollain, A. Schmid, Wirelessly powered and bi-directional data communication system with adaptive conversion chain for multisite biomedical implants over single inductive link. IEEE Trans. Biomed. Circuits Syst. **18**(3), 636–647 (2024)
39. C.-Y. Wu, S.-H. Wang, L.-Y. Tang, CMOS high-efficiency wireless battery charging system with global power control through backward data telemetry for implantable medical devices. IEEE Trans. Circuits Syst. I, Regul. Papers **67**(12), 5624–5635 (2020)
40. S.A. Mirbozorgi, H. Bahrami, M. Sawan, L.A. Rusch, B. Gosselin, A single-Chip full-duplex high speed transceiver for multi-site stimulating and recording neural implants. IEEE Trans. Biomed. Circuits Syst. **10**(3), 643–653 (2016)
41. Y. Luo, Y. Yang, H. Hong, Z. Dai, A simultaneous wireless power and data transfer system with full-duplex mode for underwater wireless sensor networks. IEEE Sensors J. **24**(8), 12570–12583 (2024)
42. H. Jung, B. Lee, Wireless power and bidirectional data transfer system for IoT and mobile devices. IEEE Trans. Ind. Electron. **69**(11), 11832–11836 (2022)
43. M.-L. Kung, K.-H. Lin, Enhanced analysis and design method of dual-band coil module for near-field wireless power transfer systems. IEEE Trans. Microw. Theory Tech. **63**(3), 821–832 (2015)

Chapter 5
System-Level Wireless Energy Management Techniques

5.1 Wireless Energy Backup Receiver

Wireless authentication tags play a crucial role in securing global supply chains by verifying the authenticity of valuable products such as pharmaceuticals, automotive parts, and sensitive electronic components [1]. Counterfeiting poses a significant threat, potentially resulting in considerable financial losses and safety concerns. Conventional authentication methods are often insufficient because sophisticated attackers can exploit various vulnerabilities, including physical attacks such as intentional power disruptions, known as power-glitch attacks [2]. To overcome these challenges, researchers have developed advanced cryptographic wireless authentication tags capable of resisting such attacks while maintaining secure communication and power-efficient operations in Fig. 5.1 [4].

This paper introduces a wireless authentication tag that significantly improves security through integration of cryptographic techniques and energy management solutions [3]. The primary innovation is the use of nonvolatile flip-flops (NVDFFs) powered by ferroelectric capacitors, allowing the tag to securely store critical cryptographic information even in scenarios where power is abruptly cut off. This approach ensures uninterrupted and secure cryptographic operations and protects against intentional power-glitch attacks designed to compromise security by disrupting the tag's power supply.

Figure 5.2a illustrates the energy backup unit (EBU), which is a critical part of the authentication tag. The EBU ensures that the data of the tag remains secure and operations uninterrupted during unexpected power interruptions. It operates in three distinct modes—charging, standby, and backup. In the charging mode, the EBU actively charges an internal backup capacitor from the received wireless power. This charging utilizes a voltage doubler which effectively increases the voltage received from the wireless source to adequately charge the backup capacitor to its designated

© The Author(s), under exclusive license to Springer Nature   91
Switzerland AG 2025
B. Lee et al., *Energy-management Integrated Circuit Design for Wireless Power and Data Transfer Applications*, Analog Circuits and Signal Processing,
https://doi.org/10.1007/978-3-032-00745-2_5

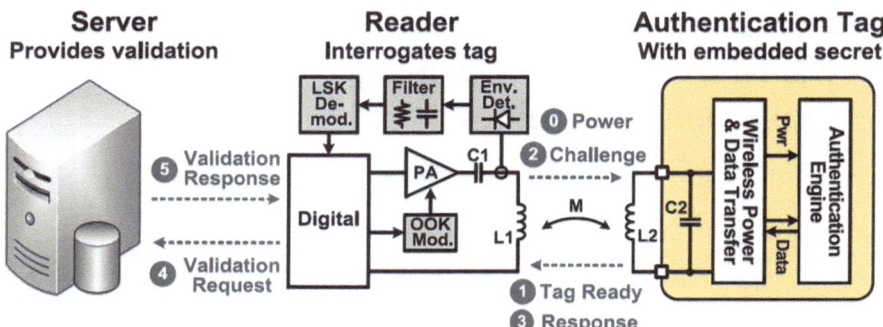

Fig. 5.1 Authentication system consisting of the tag, the handheld reader, and the back-end server [3]

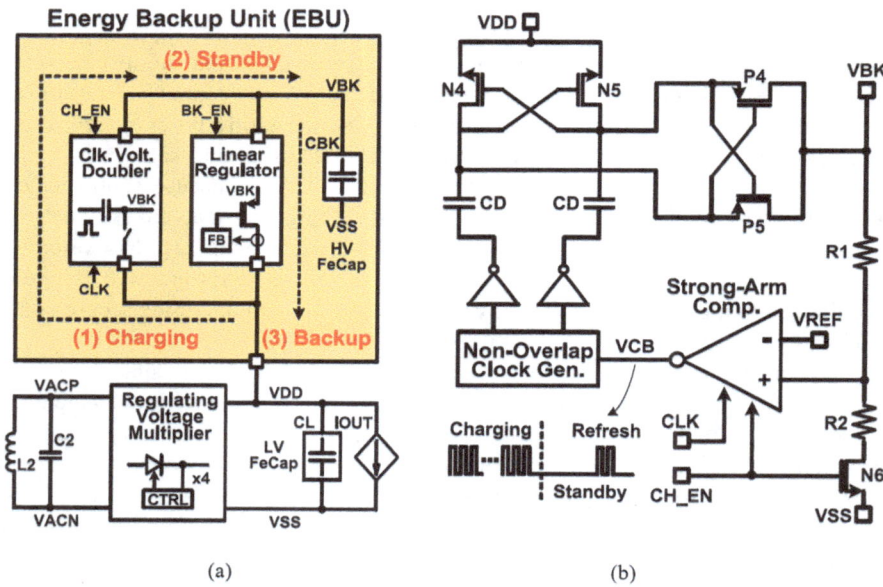

Fig. 5.2 Block diagram of (**a**) energy backup unit (EBU) and (**b**) its clock-controlled voltage doubler [3]

voltage. The standby mode is essential for maintaining the backup capacitor at a stable voltage level once fully charged, ensuring minimal leakage and readiness to provide backup power whenever necessary. In this mode, the voltage doubler periodically activates just enough to maintain the capacitor's charge without significant power consumption, making it energy efficient. When the tag experiences an abrupt loss of external wireless power possibly due to a deliberate attack or incidental disruption, the EBU seamlessly transitions into backup mode. In this mode, the previously stored energy in the backup capacitor is regulated and supplied through an integrated linear regulator to power the cryptographic processing units and NVDFFs. This backup power enables the tag to securely save the cryptographic states necessary for operation continuity.

The detailed operation of the voltage doubler within the EBU is also depicted in Fig. 5.2b. It includes a comparator circuit that constantly monitors the voltage of the backup capacitor. If the capacitor voltage drops below designated voltage, the comparator triggers the voltage doubler to recharge the capacitor to its optimal voltage level. This automated recharging mechanism ensures consistent readiness and effective energy management without manual intervention or significant power overhead.

Figure 5.3 presents the performance metrics of the RVM in terms of its voltage regulation and energy conversion efficiency. The system is capable of generating a stable 1.5 V supply (VDD) with an output power of 60 μW when the input AC signal at 433 MHz has a peak amplitude exceeding 550 mV. In addition, the tag utilizes a coil measuring 8 mm in diameter with an inductance of 35 nH, while wireless power is delivered across a 5 mm gap using a 10 mm-diameter transmitter coil possessing an inductance of 23.5 nH. This results in a voltage conversion ratio (VCR), defined as VDD divided by the input peak voltage, reaching as high as 2.73. Simulation results show that the RVM can achieve a peak power conversion efficiency (PCE) of up to 60%. However, as the input amplitude increases, PCE begins to decline due to enhanced losses across the internal regulation switches. The regulator also demonstrates strong stability, with static line regulation measured at 0.9% as $V_{AC,peak}$ varies between 0.55 and 1.2 V, and load regulation of 1.1% observed across a load current range from 10 to 200 μA. These results highlight the effectiveness of RVM in delivering both high voltage gain and reliable regulation under varying input and load conditions.

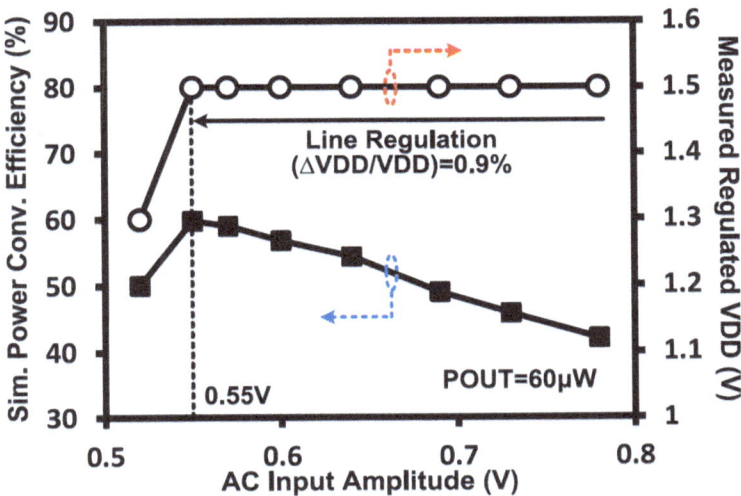

Fig. 5.3 Power/voltage conversion efficiency and regulation capability of the RVM [3]

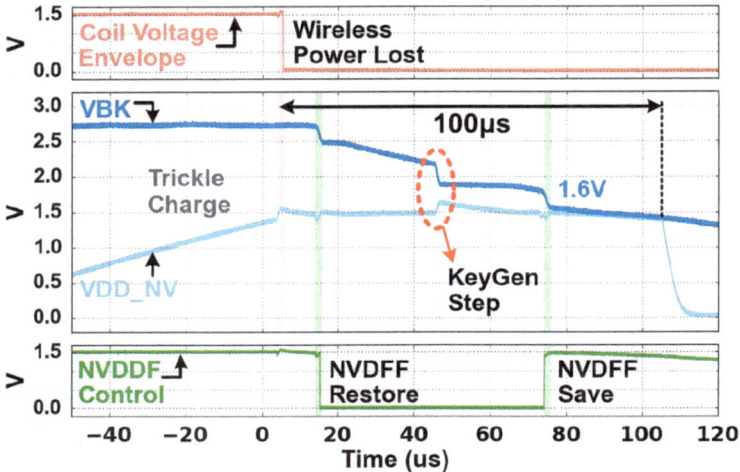

Fig. 5.4 Measured waveforms showing safe shutdown with the EBU during a worst case power interruption event [3]

The ability of the EBU to perform a reliable shutdown during a sudden power loss was confirmed under worst-case interruption conditions as illustrated in Fig. 5.4. Wireless power was cut immediately after VDDNV reached 1.5 V, but before the NVDFF restoration process began. In response, the WPDT modules transitioned into a low-power state, and the backup capacitor (CBK) provided the necessary energy to the AE for executing a key update and safely storing the data to the NVDFFs—all within a 70 μs window. Although VBK gradually declined from its initial 2.75 V, it remained above 1.5 V throughout the event, demonstrating that the backup energy was adequate. Following the completion of this sequence, the supply control logic triggered a rapid discharge of VDDNV, ensuring proper functionality of the nonvolatile flip-flops.

5.2   Energy-Reuse Wireless Power and Data Transfer

Most medical implants require a highly efficient wireless power and data transfer system, capable of transferring both energy and data wirelessly at a frequency of 13.56 MHz. Initially, the paper illustrates conventional wireless power and data transfer systems through Fig. 5.5 [6]. A conventional system typically includes three main parts: an external unit outside the body, an inductive link formed by two coils (primary coil outside and secondary coil implanted inside), and the implanted medical device itself. The external unit sends energy wirelessly through magnetic fields created by an external coil. The internal coil receives this magnetic energy, converts it into electrical energy, and powers the implant. Additionally, data from the implant (like recorded physiological signals) can be sent back to the external

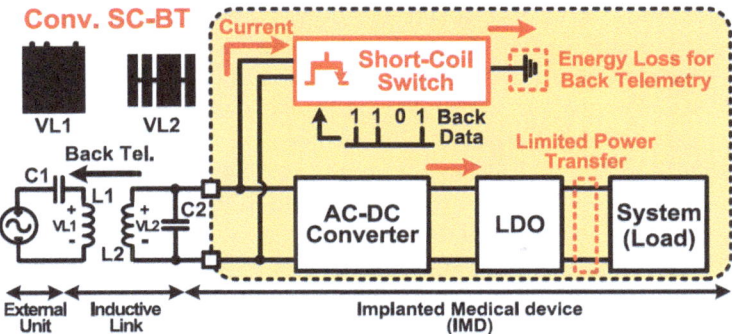

Fig. 5.5 Conceptual block diagram of the WPDT system with the conventional short-coil back telemetry (SC-BT) [5]

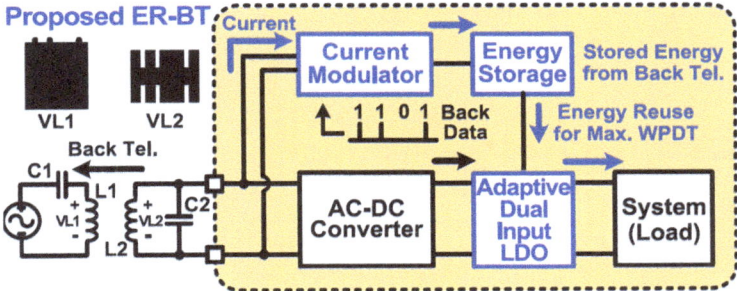

Fig. 5.6 Conceptual block diagram of the WPDT system with the proposed Energy-Reuse Back Telemetry (ER-BT) [5]

device. Traditionally, this data is transmitted by shorting the internal coil temporarily, a method called short-coil back telemetry (SC-BT) [7]. However, this conventional method has significant drawbacks: it wastes energy, as the shorting mechanism dissipates energy instead of storing it, and it prevents simultaneous data reception from outside during this transmission phase.

To solve these challenges, the paper proposes an energy-reuse back telemetry (ER-BT), as shown in Fig. 5.6 [5]. In this approach, instead of shorting the coil directly to the ground, the internal device temporarily redirects the coil current into an energy storage unit during data transmission. This stored energy is not lost; instead, it is later used to power the implant itself, significantly enhancing efficiency. Additionally, because the coil is not fully shorted, external data signals can still be detected simultaneously during this energy storage phase. Thus, ER-BT allows continuous data reception from the external unit even while the implant is transmitting data. The key innovation of this system is its ability to reuse energy from its own data transmission, greatly improving overall efficiency and reducing power waste, which is critical for safety and reliability in medical implants.

Figure 5.7 provides detailed architecture of the proposed system that implements the energy-reuse strategy. This comprehensive system includes several critical components. Firstly, two current modulators direct the coil current to an energy storage capacitor, based on the transmitted data. These modulators precisely control when energy is captured and stored, thereby encoding the transmitted information as small current fluctuations detectable externally. Next, a voltage booster elevates the stored energy voltage to an optimal level, ensuring it can effectively power the implant. Crucially, the system features an adaptive dual-input voltage regulator, which manages energy usage from two sources: the wireless energy received via the main rectifier and the energy stored during data transmissions. This regulator decides how much energy to use from each source based on the implant's power needs and the available energy stored, maximizing efficiency.

The proposed adaptive dual-input voltage regulator is shown in Fig. 5.8. This proposed regulator includes two subunits: the main low-dropout regulator (main LDO) powered directly by the main rectifier and a secondary LDO (sub LDO) that draws from the stored telemetry energy. The main LDO ensures a stable power

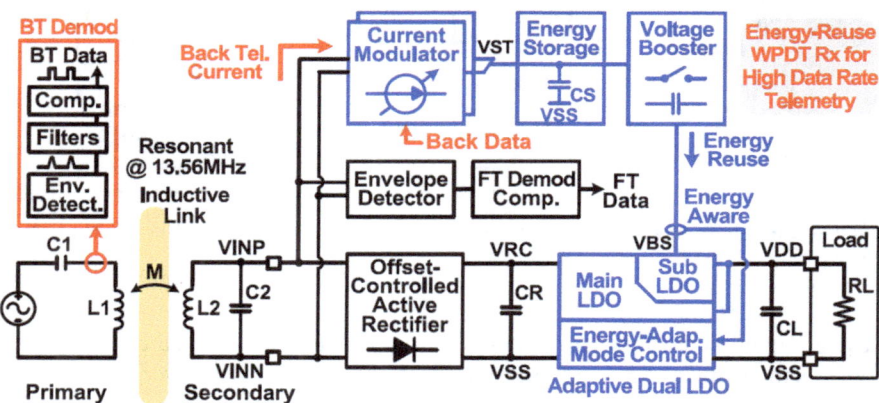

Fig. 5.7 Overall block diagram of the proposed WPDT system with emphasis on ER-BT and energy-adaptive dual-input voltage regulation [5]

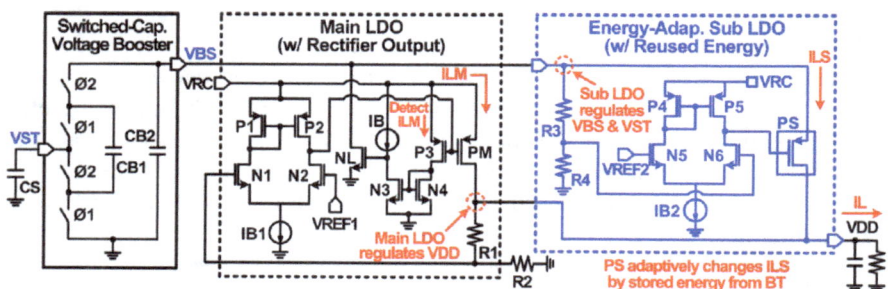

Fig. 5.8 Schematic of the energy-adaptive dual-input LDO [5]

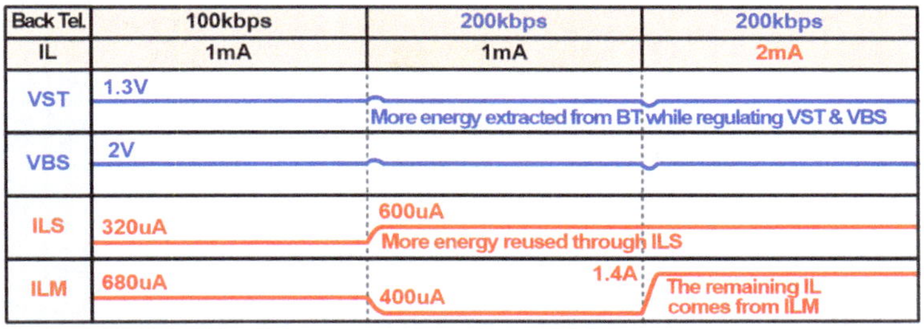

Back Tel.	100kbps	200kbps	200kbps
IL	1mA	1mA	2mA
VST	1.3V	More energy extracted from BT while regulating VST & VBS	
VBS	2V		
ILS	320uA	600uA More energy reused through ILS	
ILM	680uA	400uA	1.4A The remaining IL comes from ILM

Fig. 5.9 Operation diagram of the energy-adaptive dual-input LDO with different BT data rates and load conditions [5]

supply to the implant under normal conditions. However, during back data transmissions, the sub LDO automatically utilizes the extra energy stored from the ER-BT method to partially or fully supply the implant's power needs. It also includes a detection mechanism that avoids overcharging the implant when excess energy is available, ensuring safety and reliability. Experimental measurements presented in the paper confirm the practical benefits of this new system. At a data transmission rate of 500 kbps, the ER-BT system effectively reuses up to 42% of the energy that would otherwise be lost using traditional methods. This improved efficiency directly translates to lower overall power consumption, longer operating life, and safer temperatures within the medical implant. In conclusion, this wireless power and data transfer system significantly advances implantable device technology. By intelligently reusing energy during data transmission and adaptively managing power from two sources, it effectively increases the PCE during back telemetry.

Figure 5.9 illustrates the functional behavior of the energy-adaptive dual-input LDO across various BT data rates and load conditions. As the BT data rate increases from 500 kb/s to 1 Mb/s, more energy becomes available from the ER-BT path. This allows the LDO to draw a greater portion of the load current (I_L) from the secondary supply path (I_{LS}), utilizing stored energy from capacitor C_S and improving efficiency. The remainder of I_L is provided by the primary path (I_{LM}) through the main LDO. While doubling the BT data rate theoretically allows I_{LS} to double, practical effects—such as reduced V_{IN} amplitude and lower rectifier power due to lighter loading—slightly limit this gain. When I_L increases beyond what I_{LS} can support, I_{LM} compensates by providing the difference. However, if I_{LS} nearly meets I_L, I_{LM} may fall below a critical threshold. To maintain regulation, a monitoring circuit (I_{LM} detector) engages an auxiliary path to dissipate excess energy from V_{BS}, thereby stabilizing I_{LM} and preventing over-delivery from C_S. This adaptive control ensures system efficiency remains high, particularly when reused BT energy forms the primary supply, minimizing the load on the rectifier and reducing unnecessary power loss typical in conventional BT-based systems.

Figure 5.10 presents a comparative evaluation of the energy usage of WPDT system under different BT techniques. To maintain fairness across cases, the output power of transmitter was adjusted so that the rectifier voltage (V_{RC}) remained constant at 2 V. For simplification, the output current from the rectifier was assumed to

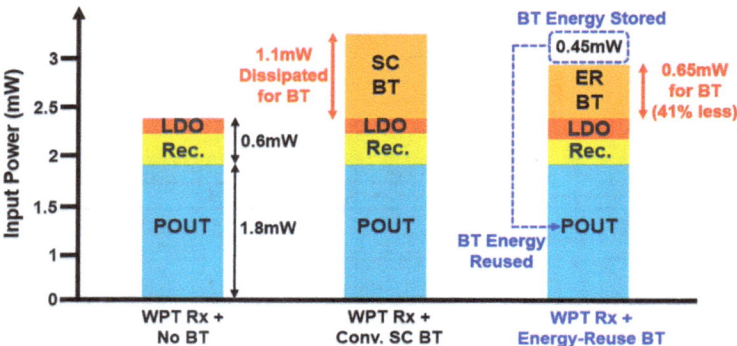

Fig. 5.10 Power consumption analysis of the WPDT Rx depending on BT schemes: SC-BT and ER-BT [5]

match that of the LDO, given the negligible internal consumption in LDO is under the microampere range. In the absence of any BT scheme, the receiver—which includes the rectifier and LDO—requires 2.4 mW of input power to deliver 1.8 mW to the load (based on $V_{DD} = 1.8$ V, $I_L = 1$ mA), translating to a power efficiency of 75%. When employing a SC-BT operating at 500 kb/s with a 0.4-µs pulse width, the BT switching circuitry adds 1.1 mW of overhead power, leading to a 46% increase in total input power. In contrast, the proposed ER-BT method reduces the input requirement to 3.05 mW—0.47 mW less than that of SC-BT. During ER-BT operation, approximately 0.45 mW of energy is temporarily stored in capacitor C_S at a storage voltage (VST) of 1.3 V, and later reused through the current modulation management (CM) unit and voltage booster (VB) path. This recovered energy, amounting to 0.31 mW, supplies about 17% of the total load power. Notably, during SC-BT operation, rectifier consumption slightly larger 0.42 mW compared to 0.4 mW (no BT) due to periods when V_{IN} drops to 0 V, disabling rectification. Conversely, ER-BT lowers the burden of the rectifier to 0.36 mW by offloading part of the load current through the sub LDO via the CM and VB, thus improving the overall system efficiency while maintaining the same 500 kb/s BT rate.

5.3 Adaptive Supply Wireless Stimulation System

Deep brain stimulation (DBS) has emerged as an effective treatment method for various neurological conditions such as Parkinson's disease, tremors, and dystonia [8]. DBS involves surgically implanting electrodes into specific regions of the brain to deliver precise electrical pulses, regulating abnormal neural activity [9]. This section presents a highly power-efficient wireless DBS system featuring adaptive power control technology designed to optimize power consumption and maintain safety during stimulation as illustrated in Fig. 5.11. The adaptive power supply system dynamically adjusts its power supply voltage based on the requirements of the

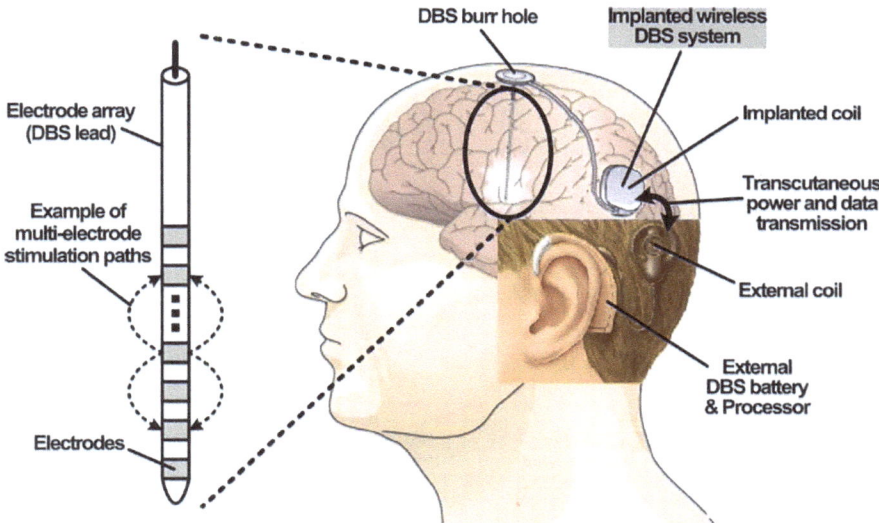

Fig. 5.11 Conceptual configuration of a head-mounted inductively powered DBS system in which power and data are transferred through the inductive link [10]

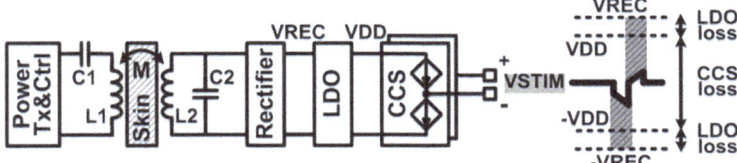

Fig. 5.12 Conceptual inductively powered stimulating system with the conventional rectifier and regulator [10]

stimulation site, enhancing overall efficiency and reducing unnecessary energy consumption.

As shown in Fig. 5.12, conventional current-controlled stimulators (CCSs) powered by inductive links typically adopt a straightforward architecture consisting of a rectifier that transforms the received AC voltage into DC voltage, followed by a low-dropout regulator (LDO) that provides a stable output voltage [11]. While this prototype is easy to implement, it suffers from significant power inefficiency. A considerable portion of the input energy is lost across the LDO and the current source. This inefficiency becomes higher when the minimum voltage is required to sustain a constant stimulation current decrease, resulting in increased power dissipation in both LDO and current source in stimulation.

To address these inefficiencies, Fig. 5.13 introduces the proposed adaptive rectifier system [10]. This advanced design incorporates an internal closed-loop control that continuously monitors and adjusts output voltage of the rectifier, closely

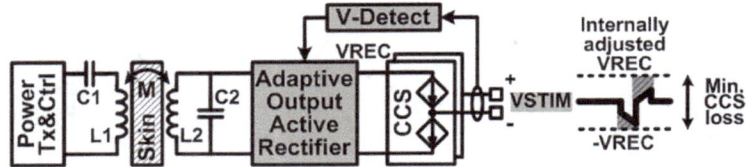

Fig. 5.13 Conceptual inductively powered stimulating system with the internal closed-loop supply control with the proposed adaptive rectifier [10]

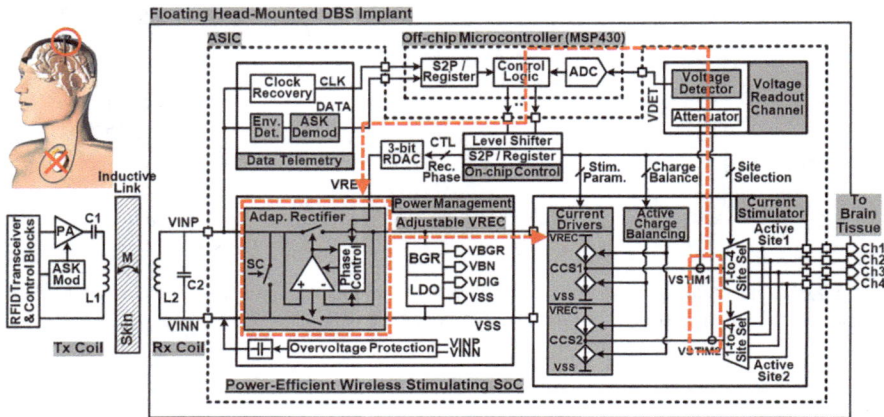

Fig. 5.14 Overall architecture of the proposed inductively powered head-mounted DBS system equipped with the adaptive supply control and the active charge balancing for both power-efficient and safe current stimulation [10]

matching the instantaneous voltage needs of the stimulation site. By adaptively controlling switching operation of the rectifier, the system dynamically sets the DC voltage level exactly as required for optimal stimulator operation. This precise control greatly reduces unnecessary voltage excess, thereby significantly improving energy efficiency. Moreover, the adaptive rectifier is designed with active synchronous switches and feedback mechanisms that substantially increase the power conversion efficiency (PCE), further enhancing system performance.

Figure 5.14 presents the system-level design of the proposed inductively powered, head-mounted DBS device. The AC power received via the inductive link is processed by the power management unit, which dynamically adjusts the rectifier output voltage (V_{REC}) based on phase control signals as described in Sect. 3.3.1. These signals are configured according to the peak voltage sensed at the stimulation electrodes, which is digitized into a reference voltage (V_{REF}) using a 3-bit resistor digital-to-analog converter (RDAC). The LDO regulator generates the digital core voltage (V_{DIG}) required for operating the control logic and low-voltage digital circuits. To prevent excessive input voltage, an overvoltage protection (OVP) module actively monitors the AC input voltages (V_{INP}, V_{INN}) and connects a detuning capacitor across the inductive input when a predefined voltage threshold is exceeded. The system incorporates

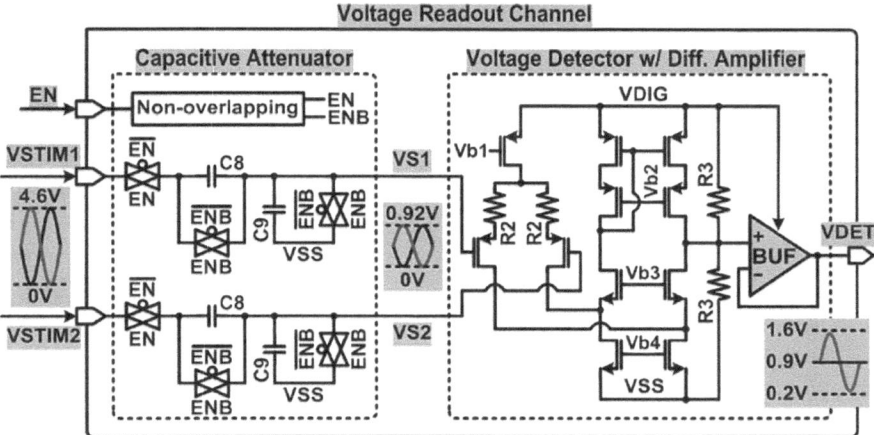

Fig. 5.15 Schematic diagram of the voltage readout channel including the capacitive attenuator and voltage detector, which are used for both adaptive supply control [10]

two current-controlled stimulation drivers (CCS_1 and CCS_2) powered directly from V_{REC}. These drivers alternately supply four stimulation electrodes, allowing high compliance operation and improving energy efficiency during stimulation. A voltage sensing circuit measures the potential difference between selected electrode sites and transmits this information to an external microcontroller (MCU), which in turn adjusts V_{REC} accordingly. This feedback loop also ensures safe operation by enabling active charge balancing, where supplementary current pulses are delivered to maintain the voltage across electrodes within a safe threshold. Communication from the external transmitter coil is decoded using amplitude shift keying (ASK), which configures the stimulation settings and electrode selection. Backward data transmission to the external unit is achieved using load shift keying (LSK), implemented by toggling short-coil (SC) switches across the secondary coil (L_2).

Figure 5.15 explains a component of this adaptive wireless DBS system—the voltage readout channel. This channel plays a pivotal role in the adaptive supply control loop by accurately measuring and generating the stimulation site voltages. The channel consists of two main parts: a capacitive attenuator and a voltage detector. The capacitive attenuator first reduces high voltages from the stimulation electrodes to safer, lower voltage levels that can be precisely measured without damaging sensitive electronic components. After attenuation, the voltage detector—a differential amplifier followed by a buffer—converts these reduced voltages into an easily interpretable output signal. The measured signal is then processed by an external microcontroller unit (MCU), which adjusts the adaptive rectifier settings accordingly. This feedback loop ensures that the voltage supplied to the stimulation electrodes consistently remains just above what is minimally necessary, optimizing efficiency and ensuring patient safety.

Figure 5.16 presents overall power efficiency from the secondary coil (L_2) to the load through adaptive regulating rectifier and stimulation. This total efficiency is

Fig. 5.16 Overall power efficiencies including rectifier and stimulator versus stimulation current [10]

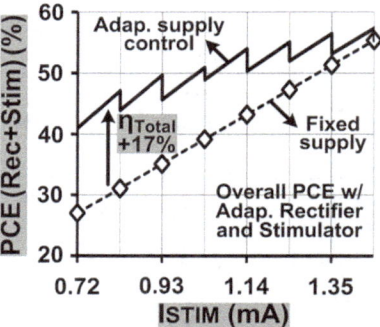

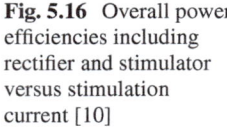

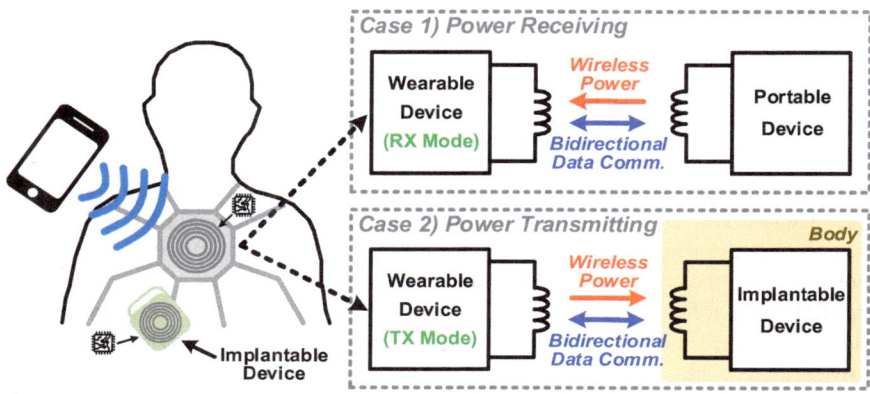

Fig. 5.17 System architecture of the proposed reconfigurable bidirectional wireless power and data transfer system [12]

calculated by multiplying the PCE of the adaptive rectifier—previously shown in Sect. 3.3.1—with the stimulation efficiency. The adaptive supply method demonstrates improved system-level efficiency, particularly under varying stimulation currents (I_{STIM}), due to its ability to maintain high rectifier performance even at lower V_{REC} levels. As a result, the total efficiency achieved with adaptive regulating rectifier ranges from 41% to 58%, outperforming the fixed supply setup, which reaches only 27–55%. This highlights the effectiveness of adaptive supply regulation in optimizing energy delivery to the load in inductively powered stimulation systems.

5.4 Reconfigurable Bidirectional Wireless Power and Data Transceiver

Figure 5.17 shows a reconfigurable transceiver IC that enables a bidirectional wireless power and data transfer (WPDT) system for wearable biomedical applications [13]. We also implemented full-duplex (FD) data transmission to facilitate real-time control and monitoring between devices. Additionally, we introduce the FSK pulse width modulation (FSK-PWM) method, which integrates data information with a clock signal for synchronized data recovery. As shown in Fig. 5.17, the proposed reconfigurable bidirectional wireless power and data transceiver (RB-WPDT) IC can operate as a Tx or Rx based on the system's requirements [12]. Furthermore, RB-WPDT IC enables FD data transmission via a single inductive link without requiring additional components such as coupled coils or antennas. The RB-WPDT IC can be applied in wearable WPDT systems. When the wearable device needs a battery recharge (in case 1), the transceiver switches to Rx mode to receive wireless power from a portable device. When the battery of the wearable device is fully recharged (in case 2), the transceiver operates in Tx mode and transmits wireless power and data to the wearable or implantable devices.

Figure 5.18 shows the overall block diagram of the RB-WPDT system. The proposed RB-WPDT IC is reconfigurable, operating as a Tx or Rx within the bidirectional WPDT system. This transceiver can be reconfigured as a differential class-D power amplifier (PA) and a full-wave rectifier depending on the mode. In Tx mode, the transceiver is configured with a PA, generating AC energy for WPT with gates of the PMOS and NMOS power transistors driven by an FSK-PWM modulator and buffers.

In Rx mode, the transceiver is configured as a full-wave rectifier, converting the AC voltage (V_{RX1} and V_{RX2}) transmitted via an inductive link into a DC voltage (V_{OUT}). The gates of the NMOS power transistors are grounded through a buffer and operate as diodes, while the cross-connected PMOS power transistors are automatically driven by V_{RX1} and V_{RX2}, eliminating the need for additional complex controllers. This single integrated transceiver can reduce the silicon area eliminating the need for separate PAs and rectifiers, resulting in a more compact system design.

In Tx mode, the transceiver simultaneously transmits wireless power and FSK-PWM downlink data by modulating the carrier signal with an FSK-PWM modulator. It also receives uplink data through an LSK demodulator. In Rx mode, the transceiver receives wireless power from a full-wave rectifier while demodulating FSK-PWM downlink data. For LSK uplink data transmission, the Rx mode

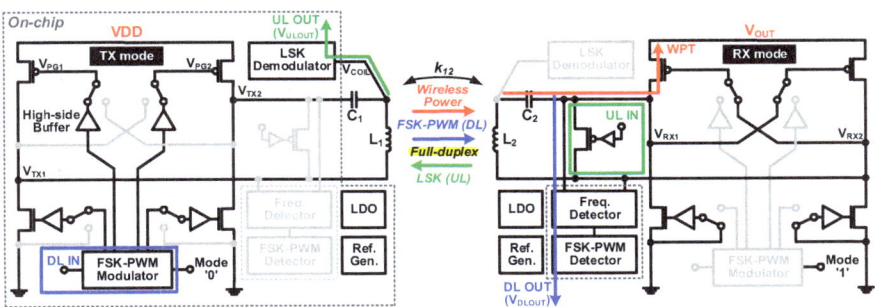

Fig. 5.18 Overall block diagram of the proposed reconfigurable bidirectional wireless power and data transfer system [12]

transceiver uses a PMOS transistor switch. The LSK demodulator consists of an envelope detector and a low-pass filter to extract the LC-tank voltage change at the Tx caused by the LSK data transmission. Since the proposed RB-WPDT IC is implemented on-chip, the overall system size can be reduced, which is especially beneficial for size-constrained wearable biomedical devices. Furthermore, the on-chip implementation enables a low-power design compared to the COTS implementation.

Figure 5.19 presents the system architecture and timing diagram of the proposed FD data transmission. In Tx mode, the transceiver transmits both wireless power and FSK-PWM modulated downlink data. In Rx mode, the transceiver receives power and transmits LSK modulated uplink data. The proposed FD method enables simultaneous transmission of FSK-PWM downlink data and LSK uplink data. The LSK method is utilized for the uplink data transmission due to its low-power consumption. Compared to other data modulation methods such as ASK and PSK, the FSK is more robust to self-interference caused by LSK uplink transmission, which is advantageous for FD data transmission. From a data demodulation perspective, the FSK demodulator detects the change in the carrier frequency, while the LSK demodulator detects the change in Tx coil voltage (V_{COIL}). This distinction in demodulation methods helps to minimize interference between them. Furthermore,

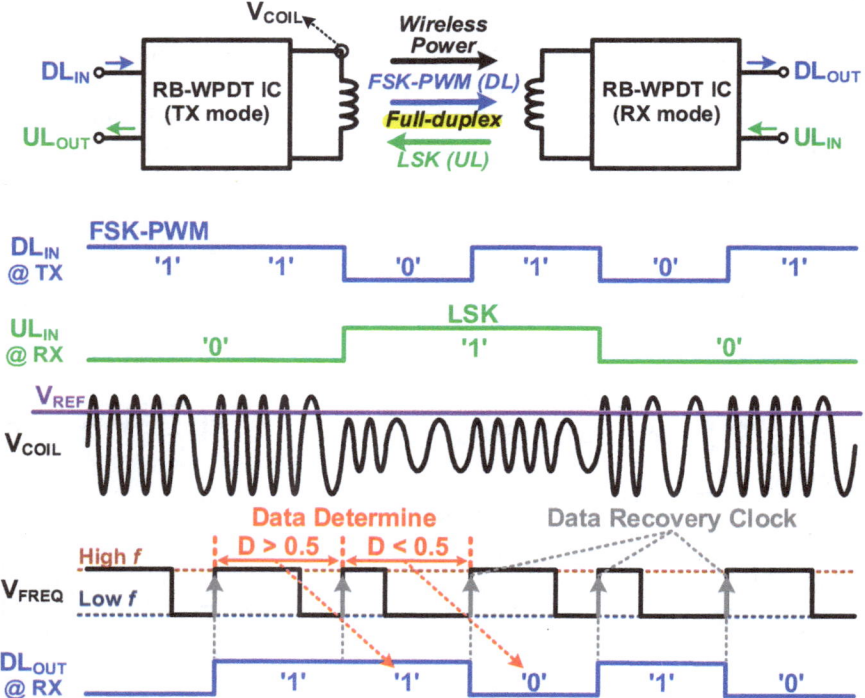

Fig. 5.19 System architecture and timing diagram of proposed FD data transmission [12]

the FSK method is beneficial for WPT systems because it enables uninterrupted power transmission along with data transmission, resulting in stable power delivery.

Figure 5.20 shows the measured waveforms of the proposed FD data transmission. In this FD data transmission, FSK-PWM downlink data and LSK uplink data are transmitted simultaneously via a single inductive link. The recovered FSK-PWM data signal (V_{DLOUT}) represents data "1" when the duty cycle of the V_{FREQ} exceeds 0.5, and data "0" when the duty cycle is below 0.5. The LSK uplink data is demodulated by extracting the voltage of the Tx coil through an envelope detector and a low-pass filter. The LSK demodulated signal, V_{ULOUT}, varies in response to the transmitted LSK data bits. The measured data rates of FSK-PWM downlink and LSK uplink are 250 kb/s and 67 kb/s, respectively. An experiment was conducted at a coil-to-coil distance of 5 mm, transmitting 250 kb/s downlink data and 67 kb/s uplink data simultaneously.

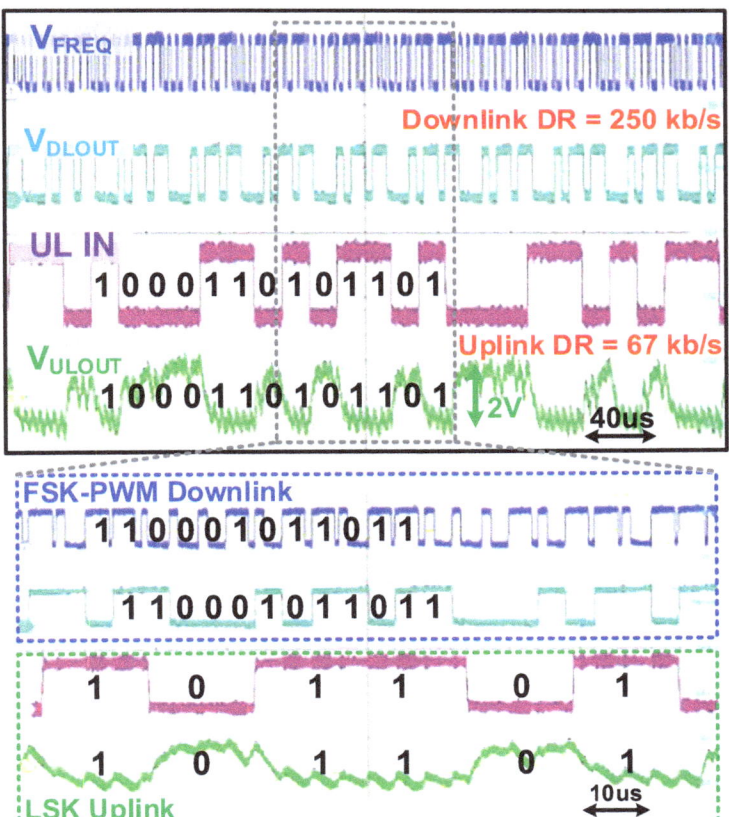

Fig. 5.20 Measured waveforms of the proposed FD data transmission utilizing the FSK-PWM and LSK methods [12]

5.5 Conclusions

This chapter presented advanced system-level techniques for efficient and secure wireless energy management in biomedical and authentication applications. Innovations such as the energy backup receiver with nonvolatile memory, energy-reuse back telemetry (ER-BT), and adaptive supply wireless stimulation significantly improve power efficiency, reliability, and functional safety. Furthermore, the reconfigurable bidirectional wireless transceiver enables compact, full-duplex power and data communication through a single inductive link. Collectively, these techniques represent critical advancements toward future miniaturized, high-performance wireless medical and security devices.

References

1. K. Zhao, L. Ge, A survey on the internet of things security, in *Proc. Int. Conf. Comput. Intell. Security (CIS)*, (IEEE, 2013), pp. 663–667
2. K. Yang, X. Jia, R. Renesse, Secure and verifiable policy update outsourcing for big data access control in the cloud. IEEE Trans Parallel Distrib Syst **26**(12), 3461–3470 (2015)
3. H.-M. Lee, C.S. Juvekar, J. Kwong, A.P. Chandrakasan, A nonvolatile flip-flop-enabled cryptographic wireless authentication tag with per-query key update and power-glitch attack countermeasures. IEEE J. Solid State Circuits **52**(1), 272–283 (2017)
4. S. Mangard, E. Oswald, T. Popp, *Power Analysis Attacks: Revealing the Secrets of Smart Cards* (Springer, New York, NY, 2007)
5. M. Kim, H.-S. Lee, J. Ahn, H.-M. Lee, A 13.56-MHz wireless power and data transfer system with current-modulated energy-reuse back telemetry and energy-adaptive voltage regulation. IEEE J. Solid State Circuits **58**(2), 400–410 (2023)
6. M. Zargham, P.G. Gulak, Maximum achievable efficiency in nearfield coupled power-transfer systems. IEEE Trans. Biomed. Circuits Syst. **6**(3), 228–245 (2012)
7. C.-H. Cheng et al., A fully integrated 16-channel closed-loop neural-prosthetic CMOS SoC with wireless power and bidirectional data telemetry for real-time efficient human epileptic seizure control. IEEE J. Solid State Circuits **53**(11), 3314–3326 (2018)
8. A.M. Kuncel, W.M. Grill, Selection of stimulus parameters for deep brain stimulation. Clin. Neurophysiol. **115**(11), 2431–2441 (2004)
9. K. Eom, H.-S. Lee, S.-B. Ku, J. Kang, H. Jung, T. Kim, J. Joo, T. Kim, Y.-M. Shon, H.-M. Lee, A 10-V-tolerant dual-mode neural stimulation system with self-sustaining dynamic supply and error-resilient digital stimulus odometer. IEEE J. Solid-State Circuits (2025). https://doi.org/10.1109/JSSC.2025.3542022
10. H.-M. Lee, H. Park, M. Ghovanloo, A power-efficient wireless system with adaptive supply control for deep brain stimulation. IEEE J. Solid State Circuits **48**(9), 2203–2216 (2013)
11. K. Chen, Z. Yang, L. Hoang, J. Weiland, M. Humayun, W. Liu, An integrated 256-channel epiretinal prosthesis. IEEE J. Solid State Circuits **45**(9), 1946–1956 (2010)
12. J. Lee, Y. Kim, D. Kang, I. Song, B. Lee, A reconfigurable bidirectional wireless power and full-duplex data transceiver IC for wearable biomedical applications. IEEE Trans. Biomed. Circuits Syst. (2024). https://doi.org/10.1109/TBCAS.2024.3483950
13. Y.K. Lo, K. Chen, P. Gad, W. Liu, A fully-integrated high-compliance voltage SoC for epiretinal and neural prostheses. IEEE Trans. Biomed. Circuits Syst. **7**(6), 761–772 (2013)

Chapter 6
Summary

Chapter 2 establishes the fundamental principles and circuit modeling of wireless power transfer (WPT) systems, particularly those utilizing inductive links. It details magnetic flux phenomena, key parameters like coupling coefficient (k) and mutual inductance (M), and presents equivalent circuit models for 2-coil links, optimizing power transfer efficiency (PTE) and power delivered to load (PDL). The chapter highlights that Rx LC-tank configurations (series/parallel) depend on Q-factor and desired voltage-current characteristics. Furthermore, it introduces multi-coil systems (3-coil, 4-coil) for enhanced efficiency and flexibility in weak coupling conditions. The chapter also analyzes essential DC-AC (power amplifiers) and AC-DC converters (rectifiers), their principles, design considerations, and various configurations. This comprehensive content provides the foundational understanding necessary for WPT system design.

Chapter 3 presented key advancements in rectifier and energy receiver design for WPT systems, focusing on improving PCE and system integration. Starting with passive rectifiers, it examined how threshold voltage drop impacts performance and how cross-coupled and active diode techniques can mitigate this. Active rectifiers, voltage doublers, and multipliers were then introduced, highlighting the role of high-speed comparators and offset control in reducing conduction loss. Multi-output resonant regulating rectifier (R^3), including dual- and triple-output designs, enables efficient power distribution across varying loads. Lastly, adaptive and resonant-mode energy receivers—such as time-interleaved and non-residual designs—offer improved efficiency and energy utilization, especially in long-distance and high-frequency applications. Together, these innovations provide a foundation for efficient, compact, and adaptable power management in modern WPT systems.

Chapter 4 focuses on efficient wireless power and data transmission techniques in inductive link-based systems. It highlights near-field data telemetry's suitability for compact wireless applications due to lower power consumption and simpler

B. Lee et al., *Energy-management Integrated Circuit Design for Wireless Power and Data Transfer Applications*, Analog Circuits and Signal Processing, https://doi.org/10.1007/978-3-032-00745-2_6

circuit designs. The chapter categorizes data telemetry by transmission methods (downlink, uplink, bidirectional) and various modulation techniques (single-carrier, multi-carrier, pulse-based, harmonic-based), discussing their characteristics. Optimized WPDT systems require balancing factors like data rate, distance, robustness, and energy efficiency. Addressing challenges in power and size-constrained applications, recent WPDT systems integrate power supply with bidirectional data telemetry via inductive links, minimizing additional components and power consumption. Therefore, designers must carefully select the appropriate system, considering practical limitations and available resources.

Chapter 5 described system-level approaches for optimizing wireless energy management in secure and biomedical applications. First, an energy backup receiver using nonvolatile flip-flops and a voltage-doubler-based backup unit was presented to ensure secure data retention during unexpected power loss, protecting against power-glitch attacks. Next, the optimal energy-reuse receiver was proposed to improve energy efficiency in medical implants. By capturing energy during back telemetry and reusing it via an adaptive dual-input regulator, the system reduced waste and enabled simultaneous power reception and data transmission. Finally, the chapter explored an adaptive wireless power system for deep brain stimulation (DBS), featuring real-time voltage regulation based on feedback from a voltage readout channel. This adaptive control minimized excess energy use while maintaining safe and efficient stimulation. Collectively, these solutions enhance wireless power transfer systems through intelligent, application-specific energy management.

Index

© The Editor(s) (if applicable) and The Author(s), under exclusive license to   109
Springer Nature Switzerland AG 2025
B. Lee et al., *Energy-management Integrated Circuit Design for Wireless Power
and Data Transfer Applications*, Analog Circuits and Signal Processing,
https://doi.org/10.1007/978-3-032-00745-2

FSC
www.fsc.org
MIX
Papier | Fördert
gute Waldnutzung
FSC® C083411

Zeitfracht Medien GmbH
Ferdinand-Jühlke-Straße 7
99095 Erfurt, Deutschland
produktsicherheit@kolibri360.de